Buzz!

Most of us crave new experiences and sensations. Whether it's our attraction to that new burger place or the latest gadget, newness tugs at us. But what about those who can't seem to get enough? They jump out of planes, climb skyscrapers, and will eat anything (even poisonous pufferfish)... Prompting others to ask 'what's wrong' with them. These are sensation-seekers and they crave intense experiences, despite physical or social risk. They don't have a death wish, but seemingly a need for an adrenaline rush, no matter what.

Buzz! describes the world of the high sensation-seeking personality in a way that we can all understand. It explores the lifestyle, psychology, and neuroscience behind adrenaline junkies and daredevils. This tendency, or compulsion, has a role in our culture. But where is the line between healthy and unhealthy thrill-seeking? The minds of these adventurers are explained page by page.

Kenneth Carter is a board certified clinical psychologist and the Charles Howard Candler Professor of Psychology at Oxford College of Emory University, USA. Previously, he worked as a senior assistant research scientist in the prestigious Epidemic Intelligence Service of the Centers for Disease Control and Prevention, where he researched smoking as a risk marker for suicidal behaviors in adolescents. He has had a long-standing interest in the psychology of thrill-seeking and has published extensively on the subject.

Buzz!

Inside the Minds of Thrill-Seekers, Daredevils, and Adrenaline Junkies

KENNETH CARTER
Oxford College of Emory University

CAMBRIDGE
UNIVERSITY PRESS

University Printing House, Cambridge CB2 8BS, United Kingdom

One Liberty Plaza, 20th Floor, New York, NY 10006, USA

477 Williamstown Road, Port Melbourne, VIC 3207, Australia

314–321, 3rd Floor, Plot 3, Splendor Forum, Jasola District Centre,
New Delhi – 110025, India

79 Anson Road, #06–04/06, Singapore 079906

Cambridge University Press is part of the University of Cambridge.

It furthers the University's mission by disseminating knowledge in the pursuit of education, learning, and research at the highest international levels of excellence.

www.cambridge.org
Information on this title: www.cambridge.org/9781108738101
DOI: 10.1017/9781108557252

First published 2019

Printed in the United Kingdom by TJ International Ltd. Padstow Cornwall

A catalogue record for this publication is available from the British Library.

Library of Congress Cataloging-in-Publication Data
Names: Carter, Kenneth, 1967– author.
Title: Inside the minds of thrill-seekers, daredevils, and adrenaline junkies / Kenneth Carter, Oxford College of Emory University.
Description: Cambridge, United Kingdom ; New York, NY : Cambridge University Press, 2019. | Includes bibliographical references and index.
Identifiers: LCCN 2019018394 | ISBN 9781108738101 (alk. paper)
Subjects: LCSH: Sensation seeking. | Risk-taking (Psychology)
Classification: LCC BF698. 35.S45 C37 2019 | DDC 152.4–dc23
LC record available at https://lccn.loc.gov/2019018394

ISBN 978-1-108-73810-1 Paperback

To all those who bring delicious chaos into my life.

CONTENTS

PREFACE: MY FASCINATION WITH THRILL

Do you ever wonder how two people can have the same experience but react to it in completely different ways?

I do. I think about this question all the time. As a psychologist it's sort of my job.

Consider the following: Two friends are at an amusement park and decide to take a ride on the Twisted Cyclone roller coaster. Two minutes and forty-four seconds of twists, dives, and jerks later, the ride is over. When they emerge, one is terrified, the other is exhilarated. The one who is terrified is breathing hard, his legs are shaking, and his heart is pounding. He did *not* have an enjoyable experience. The other looks almost tranquil in her satisfaction and is ready to go again, except this time she wants to ride in the front car. Two people, same situation, totally opposite experiences. Why does this happen? How can two people have such completely divergent responses to the same stimuli?

Of course, you can say, "They are different." But how are they different? Why are they different? What makes them so different?

These are the kinds of questions that psychologists bump up against all the time. On the one hand, I'm quite aware that people are unique – each of us has our own physiological, psychological, and cultural influences and predispositions that drive what we think and what we do. On the other hand, there are patterns to these thoughts and behaviors. Psychologists are always looking for ways to describe, explain, and even predict what people do; it's in the job description. I'm looking for the patterns in the seemingly unpredictable mélange of human behavior. My job is to find these patterns not only to understand people, but also to help them understand each other and even themselves. If you look carefully, you can see the patterns too.

I'm that guy who hates roller coasters. They aren't fun for me. I've ridden them many (many) times with friends, so it's not simply a matter of exposure. Sure, I can tolerate them and I ride when I'm pressed, but it's not fun. How someone could get off a roller coaster and not only be ready to ride again, but also actually feel happier, even more tranquil *after* the ride, has baffled me for years.

What's more, I am a person who is relatively staid. I'm a professional academic who spends the vast majority of his time in the library, behind the computer, or in front of students lecturing. My life is ordered. I wake up at the same time and go to sleep at the same time pretty much every day. I don't eat exotic foods. I don't seek out new experiences. I am perfectly content quietly doing my work. I bask in the subtleties of experience and my predictable day is a luxury. I crave calmness.

Yet, I see people who are almost perpetually and intentionally drawn not only to literal but also to metaphorical roller coasters. From the outside they seem to seek out chaos: students who change their entire course schedule the morning of the first day of classes; clients who propose marriage on the second date; friends who leave wonderful jobs to move to a different city on a whim. Some people seem to attract problems and drama like honey attracts ants. These folks often struggle to live in modern society where having a tolerance for monotony may have serious advantages.

You may have met people like this (or maybe you are like this yourself). Constantly moving from job to job, relationship to relationship, place to place. Some struggle with mental health issues. Most don't. But the underlying likeness between them is an inability to tolerate the mundane, an itch for excitement.

Take for example my friend Andrew. Andrew has an industrial-sized case of wanderlust. By the time Andrew was 27 he'd moved 13 times (to three different countries), been in nine different relationships, and had six different careers. When I asked him if moving so many times was difficult, he laughed. "No, it wasn't a challenge at all. It was an adventure."

I've met so many people like Andrew in my life that I began to wonder what they had in common. Are there people who are chaos junkies? Is there some psychological model to explain why some people are attracted to drama?

When I come up with a question like this, the first thing I do is hit the library and geek out, digging into thought pieces, research studies, articles, and books to see if someone has already answered the question I am pondering (you'll find these sources in the notes).

This is how I stumbled across the work of Marvin Zuckerman and his investigation into the high sensation-seeking personality. You'll learn more about Zuckerman's work in Chapter 1, but the essence of what he discovered is that there is a subset of people who crave stimulation and thrive in environments that would seem overstimulating, even chaotic, to the rest of us.

I became fascinated by the idea that there are people in the world who seek out stimulation and thrive on chaos. It is so contrary to my own experience that I had to learn more. I wanted to see if I could get a glimpse into what it might be like to be the kind of person who came off the roller coaster exhilarated instead of exasperated.

That was the beginning of my two-year journey to better understand the high sensation-seeking personality. I dug deep into the research, I interviewed high sensation-seekers and even followed them to the tops of cliffs and into exotic restaurants. I pried into their motivations. I asked them why they crave stimulation and questioned them about how their seemingly strange behavior impacts their daily lives, their careers, their relationships and more. Over its 70-year history, psychological research has yielded fascinating findings and constructed telling theories about sensation-seeking.

In Chapter 1, we'll examine the history of sensation-seeking and learn how various personality theories have tried to understand thrill-seekers, we'll also examine the components of sensation-seeking, and you'll take a sensation-seeking scale to learn about your own thrill (or chill) seeking personality. You'll also have a chance to examine the profiles of other sensation-seekers.

High sensation-seekers are, indeed, different than the rest of us. As you'll learn in Chapter 2, there is some evidence that they experience stress differently, their neurochemical makeup may also be different than the average person, and they may even be genetically predisposed to seek out sensation for evolutionary reasons.

In Chapter 3 we'll explore the everyday life of a high sensation-seeker. We'll delve into their preferences for jokes, travel, and even what they like to eat.

In Chapter 4 we'll dive into extreme sports and other adventures that attract high sensation-seekers and explore why many people with high sensation-seeking personalities crave activities like skydiving, rock climbing, and adventure travel. We'll look at extreme sports like tough mudder races, ice climbing, and wingsuit flying and what high sensation-seekers get out of those experiences.

We'll shift gears in Chapter 5 and consider the relationships of high sensation-seekers. Certain relationship patterns and challenges are common among people with high sensation-seeking personalities. We'll explore both friendships and romantic relationships.

In Chapter 6 we'll examine careers and work. People with high sensation-seeking personalities tend to gravitate toward certain kinds of careers. This chapter explores the often risky careers of high sensation-seeking personalities and how their personality traits can benefit and cause problems at work.

Chapter 7 looks at the dark side of sensation-seeking. For most people, high sensation-seeking isn't a problem, but for some it can be. Those with high sensation-seeking personalities sometimes struggle with problems, including anger, addiction, gambling, substance use and abuse, and crime and antisocial behavior.

And in Chapter 8 we'll try to get to the bottom of the question "Is high sensation-seeking a superpower or a super problem?" High sensation-seeking doesn't have to get you into trouble. In fact, it can be a positive force. People with high sensation-seeking personalities are active and curious, and they experience less stress and anxiety than those with low sensation-seeking personalities. While avoiding excessive risk-taking is important, high sensation-seeking can help some people get more out of life.

High sensation-seekers are different than the rest of us, but does this difference make them out of control and broken, or does it make them fascinating? Are high sensation-seekers a dangerous part of our world, or are they integral to it? Are these seekers of sensation also seekers of chaos, or is something else at play here? This book is my attempt to answer these questions. I sought to discover why high sensation-seekers do what they do, and over the course of that journey I came to see life through their eyes just a little bit. What I discovered was not at all what I predicted.

ACKNOWLEDGMENTS

There's a particular formula for being helpful: a lot of encouragement, a dash of support, and even sometimes a pinch of constructive abuse. That's where it starts. Not everyone gets this right, but I'm fortunate to have people in my life who have mastered this tricky recipe and made this project possible. A sampling of the many deserved acknowledgments follows.

This book would have consisted entirely of blank pages had it not been for the guidance and assistance of my agent, editor, and friend Cecelia A. Cancellaro of Word Creative Literary Services. From mentor to editor, to advocate to cheerleader, it's hard to overstate how valuable Cecelia has been to me. Her unflinching faith, artful strategies, and keen editing eye have made her a powerful partner. She is a literary alchemist.

Special thanks go out to Jennifer Danforth, William Wahlgran, and Spencer Smith for their help in the early days of the manuscript; to Jonna Kwiatokowski for her wonderful learning objectives and student discussion questions; and to coding wizard Jennifer McFadden for making my midnight mumblings and napkin sketches into the magic of the brief sensation-seeking survey on my website.

A hat tip to all the people I interviewed, wrote about, or quoted. There were many who weren't mentioned specifically in the book but who really helped me to think about high sensation-seeking and who transformed the tables, graphs, and statistics of research studies into the reality of how people live their lives every day. Thank you all for being so generous with your time and allowing me to look into your lives.

The team at Cambridge is truly exceptional. Thanks go to Emily Watton, Grace Morris and especially to my commissioning editor, David Repetto, whose patience is only exceeded by his encouragement and good nature.

I also want to thank the many scientists whose works are described, cited, and synthesized in this book. I'm especially thankful to Marvin Zuckerman, the father of sensation-seeking.

I'm grateful to Jack Hardy, Debbie Woog, Teddy Ottaviano, Michael McGloin, Susan Ashmore, Sharon Lewis, Jennifer McGee, Molly McGehee, Shira Miller, DJ VanCronkhite and of course my parents, Bill and Eugenia Carter. There are many other people and organizations who have been important in this process. You can find their names on my gratitude website www.drkencarter.com/thanks. Thank you all for relentless support, encouragement, and reassurance. It absolutely made a difference.

1 WHAT IS SENSATION-SEEKING

Cliff diving isn't a typical activity for anyone, much less a person who is afraid of heights. But Mike,[1] a 20-year-old intern living in Atlanta, does it as often as he can, despite this fear. He's also gone skydiving at least four times. The first time, he was a little disappointed. "I actually wasn't scared at any point, which was weird." The second time Mike actually told his guide, "The last guy failed to scare me, so I want you to scare me." Even by his own reckoning, this isn't something you should say to a skydiving coach. I couldn't help but ask, "Well, did he scare you?" "Yeah," he said, "he went about it very cleverly. Beforehand he told me that when they've got somebody who isn't cooperating (apparently some people will grab the guide's arms or something when they should be pulling the chute), they spin the person around really fast. Because this increases G-force, the person passes out and the instructor can get them safely to the ground. So we're in the middle of free fall and that is basically what he does to me. He takes my hand and bends it down slightly, so I start spinning extraordinarily fast in one direction. Then he stops me, and we spin in the other direction extremely fast. Then the next direction extremely fast, and the next. My eyes were pretty much popping out of my head . . . eventually he pulls the chute, and before I knew it we were just coasting again."

My heart started racing and I felt dizzy just hearing about this, but Mike explained all of it in a voice that seemed a little too calm for someone who insists he doesn't like heights. And his aeronautic feats don't end with skydiving and cliff diving. He also hang glides. The more terrifying, the better, and Mike seems to reserve that label for his cliff diving exploits. "As far as utter and complete fear beforehand, I'll give that to cliff diving every time . . .'cause I am quite scared of heights. Even if it's only a 35 foot jump it still gets the blood pumping quite a bit . . . Each time I am like, 'why am I up here?' I'm literally jumping from heights that I hate into the water that's not my favorite." (Mike is not a big swimmer.)

Mike is partial to spending his leisure time participating in activities that seem dangerous to the rest of us like bull runs, zorbing, and eating strange foods. As for bull runs, the Pamplona run is perhaps the most famous. Originally, its purpose was to move the bulls from the fields to the bullrings for bullfights that celebrated various festivals across Spain. Now it has become a local tradition and a worldwide phenomenon. Across the United States you can find great bull runs that are held in arenas and on racetracks. The idea? Well, you get out on the track and a pack of bulls is released behind you. You run like crazy or get trampled by a herd of cattle. Some people even take selfies along the way. It might seem like a strange way to spend your downtime, but they are actually pretty popular. Mike loves them. "Once you're in the actual run it's a sort of out of body experience, but intensely adrenaline packed." Mike has also been zorbing – a sport in which you are strapped inside a capsule that is then placed inside a gigantic transparent ball. The ball is rolled along the ground or down hills. It's like an enormous hamster ball for humans. It sounds nauseating, but apparently some people can't get enough of it. And Mike has eaten fugu (pufferfish – which numbs the tongue and lips). Although, disappointingly for him it was only the moderately poisonous type. And things he's not done yet that are on his to do list include: swimming with sharks, bungee jumping, and saving a human life.

~

Sophie can't seem to get enough out of life. Endurance athlete, blogger, marketer, model, and motivational speaker, she's a lover of challenges, determined to experience life's adventures. Sophie quit her job running the commercial division at a major UK tech startup because she was tired of being safe and was ready to live her life guided by her personal mantra, "one life, live it!"

And she's not kidding. She's packed more in the last few years than some people do in a lifetime. She's completed an adventure race through the jungle of Borneo, cycled the 244 miles from London to Paris in 24 hours, and in 2014 she was the first (and so far only) person to cycle the Alps and climb the highest mountains in all eight alpine countries. She says she enjoys her "pain cave." "I love pushing myself physically and mentally. I love being in the pain cave because it's there I find out the most interesting things about myself and that helps me to learn and grow into the person and athlete I'd love to become."

Sophie writes a fitness lifestyle and adventure blog and recently moved from Great Britain to France to prepare for her next adventure. She cycles, she climbs, she runs, she travels.[2] She tried skydiving but didn't enjoy it because she says she's not into high adrenaline activities. For Sophie, the satisfaction comes from pushing herself in order to conquer challenges. She's on a mission to inspire others to undertake extreme adventures too.

~

Kirill Vselensky loves taking pictures, especially travel photos. He shoots landscapes, buildings, bridges, landmarks, selfies, nothing too unusual – except that his shots are captured from atop some of the world's tallest buildings. Kirill loves to climb to the top of skyscrapers, bridges, anything climbable, and take pictures of himself dangling hundreds of feet above the ground – suspended without any safety gear. Known as the Russian Spider-Man, he is one of Russia's extreme climbers, called roofers, who are known to hang off buildings by only their fingers. He snaps incredible, whoa-tastic photos.[3]

Kirill should have "do not try this at home" permanently tattooed on his forehead. Roofers like Kirill sneak their way to the tops of buildings and perform tricks like standing on one leg, balancing on the side of window ledges, teetering on the edge of roofs, and climbing up huge structures without any safety rope or climbing gear. His biggest fear? It's not falling – it's "to be detained."

Although he now has over 55,000 followers on Instagram, he was seeking out sights long before anyone was watching. "As a kid I used to love to visit people as every time there was a new view from the window, it was an easy way to find adventure in your own city," Kirill explained to a newspaper reporter. Now what he sees is much more extreme. He started scaling buildings in 2008, because he "likes the views." When he was asked what goes on inside his head, he replied, "Nothing special. I just try to think about hanging tight and staying alive."[4]

~

This is a book about Mike, Sophie, and Kirill. It's about people who perform their best in highly stimulating and emotionally charged environments. It's a book about some of your friends, family members, or coworkers who fit the bill. It might be a book about you too, if you're one of those people who craves new experiences in work, in friends, and in fun. It's a book about people who base jump, spelunk, drive ambulances, and chase tornadoes. It's about thrill-seekers, adrenaline junkies, people looking for a buzz. It's about what became known in psychological circles as "the high sensation-seeking personality" or HSS for short.

If you aren't a thrill-seeker, it's entirely likely that these kinds of actions appear irrational and maybe even foolhardy. It may seem like thrill-seekers have a death wish. This is what Sigmund Freud might have believed, as you'll discover later in this chapter. It's also what I believed for a long time. In fact, it's one of the reasons I've spent so much time reading about, researching, and interviewing thrill-seekers all around the world. I began to wonder what could drive a person to intentionally seek out activities that were so utterly intense, even chaotic. Why would someone risk their life running with the bulls? Why would someone hang from a building or quit a high-paying desk job to spend more time in their "pain cave"? What drives people to seek out the most dangerous, even outrageous experiences they can find? Why would they risk swooping around in a wing suit when they could relax with a nice book on the beach? Do they really have a self-destructive urge? Is it genetic? Biochemical? Is it a modern social phenomenon? Or is something else at work here?

These are the questions we'll explore in this book. We will investigate the lifestyle, psychology, neuroscience, and environmental factors that influence people with high sensation-seeking personalities. We'll examine both the healthy and the unhealthy aspects of high sensation-seeking. We'll look at the habits and havoc this kind of personality creates. Along the way, you'll discover that high sensation-seekers' motivations and their experience of the buzz are not what most people might assume.

But what is "high sensation-seeking"? What does that term even mean?

What is High Sensation-Seeking?

To some extent, we all crave complex and new experiences – that is, we all seek new sensations. Whether it's our attraction to the new burger place down the street, the latest shiny gadget, or the newest fashion trend – newness tugs at us. It's simply human nature.

What sets the high sensation-seeking personality apart is that it craves these exotic and intense experiences, despite physical or social risk. Consider Kirill. He knows that hanging off buildings is risky (who doesn't); but he does it anyway. Is it because he's seen people on TV, in the movies, and on YouTube do this stuff? That's certainly part of it. It's true that the extreme products, activities, and entities that have emerged in the last decade – X-games, The Extreme Sports Channel, and Red Bull – responded to our collective interest in thrill-seeking as a spectacle if not a personal endeavor. And these extreme activities have spread quickly as their early adopters – people with high sensation-seeking personalities – devoured them with gusto and shared their experiences enthusiastically online. However, I don't think we are currently inspiring more thrill-seekers through TV and social media as much as these outlets are giving those who are already high sensation-seekers permission or even new ways to indulge in their passions. Why do I believe this? Because thrill-seekers have been around a long, long time, and people have been trying to understand them since the very birth of psychology.

A History of Thrill-Seeking

Bolting from bulls for fun isn't new. The first recorded running of the bulls dates back to 1591 when the people of Pamplona, fed up with the bad weather in early October (when the fiesta San Fermin was traditionally held), decided to move the celebration to July.[5] It continues to be held in early July to this day. While there are no written records that precede this, some believe that this tradition may date back to the 12th or 13th centuries. Hang gliding dates back even farther – all the way to 6th-century China where gigantic aerodynamic kites were built that allowed an average-sized person to sail in the wind.[6] This quest was pursued in earnest centuries later when the early experiments in aviation began in the late 1800s.

Thrill-seeking and watching thrill-seekers has probably been around since there was danger to be watched. From gladiator games in the ancient Roman arena, to modern mud runs, humans have had a passion for both thrill as a pastime and as a spectator sport. Over the years not much has dampened the passions of people who are captivated by living-on-the-edge sports, especially those that are death defying. Suffice it to say, high sensation-seeking has been around for a long time. The desire to understand why some people are driven to engage in these activities has been around for just as long.

Personality Theory: Trying to Explain the Thrill-Seeker

Our personality is our pattern of thinking, feeling, and behaving – the enduring nature of who we are. Personality determines not only what we like but also why we prefer one thing over another. Personality motivates what we do and how we do it. It influences our choice of friends and hobbies. If all that and more are true of our personality, then thrill-seeking must be reflected in some aspect of our personality.

Thanatos: The Death Wish

Sigmund Freud, the founder of psychoanalysis is often considered to be the founder of one of the first well-organized grand theories of personality. Freud describes the landscape of the mind as having several realms, some in awareness and some beyond. Freud suggested we have three main personality structures. I'm sure you've heard of them before: the id, ego, and the superego. The id, according to Freud, is the part of the personality that operates on the pleasure principle. What's pleasurable for the id? It's simple, the reduction of tension. This means that the id is always looking to reduce tension – mostly the tension that comes from basic physiological needs and drives. The id isn't smart, but it knows what it likes, and what it likes is reduction of tension. What kind of tension? All kinds. People have many physiological needs and drives, and when these drives are unmet, tension builds up. Being thirsty creates tension; sipping a tart lemonade will reduce that tension. Being hungry creates tension; munching on a juicy hamburger will reduce that tension. When the tension is reduced, the id feels satisfied.

Some of the physical needs that motivate the id keep us alive, cause us to eat, drink, and perpetuate the species by having sex. Freud grouped these drives into a cluster of instincts called

Eros, or the life instinct, and those Eros instincts reduce tension associated with basic biological drives.

But life instincts aren't the only things that will motivate the id; there's another way that the id seeks the pleasure principle. Freud also suggested that the id has a death instinct, an unconscious desire to be dead – the ultimate state of tension reduction called Thanatos (see? Not that smart – but it knows what it likes). Freud suggested that Thanatos is a way in which we reduce tensions that are aggressive and destructive. Think about the amount of violence in movies and even cartoons and take a look at the top movies, video games, or sports in any given year and I'll bet hitting, killing, and shouting dominate the interactions.

There was a famous exchange of letters between Sigmund Freud and Albert Einstein in 1932. Einstein wrote to Freud on behalf of the League of Nations. Einstein expressed concerns that despite our advanced civilization, war was still a reality. He wondered if Freud had insight into why. In his response Freud emphasized that both the Eros and Thanatos instincts were essential and rarely operate in isolation. In fact, it's present in every living creature.[7] When we think of thrill-seekers as "having a death wish" we are largely channeling Freud.

But do thrill-seekers have a death wish? Probably no more so than the rest of us. The thing is, when you actually talk to most thrill-seekers they don't express a death wish at all. In fact, one of the most surprising discoveries I found on my journey to understand thrill-seekers is that it's almost the opposite. Consider what Mike said when I asked him, "Is there anything that you'd like people who are not adrenaline junkies, or thrill-seekers, to know about thrill-seekers?"

He replied, "I guess it would be mostly the nots, not what they are but what they are not. They are not suicidal for one. I have no interest in dying skydiving. It's actually way safer than getting in your car and driving to work. What I have is a kind of an addiction to life for lack of a better word. I think it's not a disregard for life, but an addiction to life and trying to intensify moments instead of dull them out."

As odd as this might sound, this is what I heard from people over and over again, and it's a recurring theme throughout this book. It's totally counter to what most people might guess. In fact, it may well be true that high sensation-seekers are more in love with life than any of us could imagine.

Freud isn't the only expert who had a theory of why certain people seem to seek out complex and intense activities. His disciple Carl Jung had some thoughts on the matter, too.

From Death Wish to Extroversion: Jung's Response to Freud

Carl Jung had a slightly different idea about what drove human nature and why some people were drawn to the softer things in life while others were more outgoing and risky. Instead of Eros and Thanatos, Jung spoke about introversion and extroversion which he suggested were indicators of your preferences for the external world.[8] Introverts prefer their internal world of thoughts to the extrovert's external world of people. Introverts may be reserved when interacting socially. Extroverts, on the other hand, prefer the external world to the internal world. While some introverts tend to be outgoing and sociable, the important aspect for the introversion–extroversion dichotomy is actually energy.

Some people think that introverts are shy and keep to themselves and extroverts are outgoing, but there's more to it than that. The concepts of introversion and extroversion also include how people recharge. Introverts recharge by being by themselves while extroverts tend to recharge by being around others. One way to think about it is what you find more interesting – what's inside your head or what's outside of your head? Introverts find what's going on inside to be much more interesting. That's why they can distract themselves with their own thoughts and will retreat there for comfort and recharging. They can find others exhausting. Extroverts, on the other hand, prefer the world outside their head. They find others energizing and can get grumpy if they are alone too much.

Although it's easy to imagine that these qualities relate directly to how much sensation a person desires – low sensation-seeking individuals being introverts and high sensation-seekers being extroverts – it's not that straightforward. Research has suggested that sensation-seeking and extroversion are relatively independent.[9] This means that they are different concepts and that you can have a low sensation-seeking extrovert or a high sensation-seeking introvert. Consider the fact that there are many thrill- and adventure-seeking activities that are relatively solitary, such as rock climbing. In fact, I met many introverted thrill-seekers along the way, and you'll meet some of them in this book. So, if introversion and extroversion don't explain thrill-seekers, could it simply be a basic personality trait?

Trait Theory: Eysenck and the Big Five

The contemporary view of personality has moved away from Freud and Jung to use traits to explain personality. We use traits in our language all the time to describe the personalities of the people we know (Molly is so patient and kind). One way to think of a trait is a stable quality that differentiates one individual from another. And the trait theory of personality focuses on identifying the traits that summarize and predict a person's behavior.

One of the things that trait theorists focus on is the number and kinds of traits that you can use to describe a person's personality. Sure, you could just use every word possible in the English language to describe a person, but with nearly 14,000 available traits to choose from, you could fill a book chapter just listing someone's traits. Psychologists needed a simpler, more elegant solution. They needed to determine which traits of the thousands that exist are the most important? Hans Eysenck says you only need three.

Eysenck: The Power of Three

Hans Eysenck was a psychologist and a major contributor to the modern scientific theory of personality. He developed a distinctive three-factor model of personality structure based on three dimensions of personality: extroversion, neuroticism, and psychoticism.[10]

According to Eysenck, extroversion and introversion are created by either inhibition or excitation in the brain. Excitation is the brain waking itself up, while inhibition is the opposite, the brain calming down. Extroversion is associated with strong inhibition tendencies while the introvert has weaker inhibition.[11] According to this theory, pursuing or shunning exciting situations (such as social situations or noise) is a tactic for maintaining optimal levels of arousal. For example, researchers had subjects choose the level of background noise they preferred while working on a matching task. Introverts chose noise levels that were much lower than those of extroverts, and each group performed best under their preferred level.[12] If you think about thrill-seekers in this way, they are finding the amount of excitation that works best for them (more on this later).

While Eysenck suggested that the introversion–extroversion dimension is driven by excitation and inhibition, neuroticism, or emotional stability, is based on how easily the body's stress system is activated. Those who trip the body's stress response easily have low activation thresholds and have a harder time inhibiting their feelings. This means that minor events can make them stressed. Those with low levels of neuroticism, on the other hand, have higher activation thresholds, take longer to experience negative feelings and are harder to unnerve.

The last trait that Eysenck described is called psychoticism. Psychoticism is a trait that describes how tough minded a person might be. Despite its name, psychoticism doesn't mean that a person has psychotic tendencies, rather the trait is associated with recklessness or disregard for convention. Those who score high on psychoticism can be inflexible, creative, sometimes inconsiderate, quick to anger, and reckless. The physiological basis suggested by Eysenck for psychoticism is testosterone, with higher levels of psychoticism associated with higher levels of testosterone.[13]

Eysenck considered thrill-seeking to be a component of extroversion and impulsivity[14] which on its face makes sense. Thrill-seekers do often appear impulsive and reckless, but if you look closer some thrill-seekers are pretty methodical in their recklessness. BASE (Building, Antenna, Span, and Earth) jumps can take days to plan. What in the moment may seem a whim could possibly be the result of years of training. However, three may not be the magic number to describe thrill-seeking. Maybe it's five.

The Big Five: Is Sensation-Seeking One of Five Primary Personality Traits?

Developed by Paul Costa and Robert McCrae and nicknamed The Big Five, this theory of personality holds that there are five main personality traits: openness to experience, consciousness, agreeableness, neuroticism, and extroversion.[15]

Openness to experience is your willingness to try new things and your affection for the new. People who score high on openness to experience are curious and imaginative. People who score low might be seen as conventional. If whenever you go to a restaurant, you always order exactly the same thing, you might score low on openness to experience.

Conscientiousness is your ability to stick to the plan and act with integrity. It describes how trustworthy a person might be. People who score high on conscientiousness are seen as organized, reliable, and punctual. Those low on conscientiousness are seen as unreliable.

Agreeableness is how empathetic and interested in other people you are. It describes how trusting you might be. People who score high on agreeableness are serious, trusting, helpful, and forgiving. They might also be seen as gullible, perhaps agreeing to cash checks for stranded royalty who reach out to them via email. Those low on agreeableness can be seen as cynical and suspicious and are always looking for ways they may get tricked.

Neuroticism describes how much you worry about things or get tied up in your own thoughts and feelings. It describes how emotional a person might be. People who score high on neuroticism are seen as worrying, nervous, and excitable. They might call you and yell, "The WORST thing just happened to me today," even when what actually happened was rather minor. Those with low scores on neuroticism can be seen as cold Vulcan-like stoics.

Extroversion updated from Jung, describes your energy focus. Those who score high in extroversion are seen as social, active, and talkative. Those who score low on extroversion (aka introverts) are often reserved and quiet.

The Big Five is one of the most researched conceptualizations of trait personality out there. But a love of thrill isn't one of the Big Five traits and it's not reflected reliably by a combination of them. While some thrill-seekers have some of the impulsivity of extroverts, many others don't. Some are drawn to new things and might score high on openness to experience, but others repeat the same thrill-seeking experiences because they find something new in the activity every time.

This may very well mean that sensation-seeking is not captured by the traits in the Big Five. Perhaps explaining the thrill-seeking personality is outside the realm of the grand personality theories of Freud, Jung, Eysenck, or the Big Five. This is where a mini theory might come into play. Rather than a broad theory that explains the whole of personality, mini theories explain one thing quite well, a special tool for a special job. Perhaps a mini theory could explain thrill-seeking in an elegant way where the grand theories have fallen short. Enter Marvin Zuckerman. Zuckerman was one of the first people to see thrill-seeking as a separate and important personality trait and do real scientific research on it. However, he found his way into this area in a most unexpected way – by studying how people reacted to no stimulus whatsoever.

The Founder of Scientific Research on High Sensation-Seeking

Despite all of the previous attempts to explain the thrill-seeking phenomenon, scientific research on sensation-seeking didn't begin until the late twentieth century, and it didn't start in the base camps of Mount Everest or on the cobbled streets of Pamplona or even the racetracks of Talladega. It began in a dark room filled with nothing – literally. Researchers weren't trying to explain mountain climbing and kayaking, or running from bulls, or race-car driving. They were trying to get to the bottom of mind control.

Shortly after the Korean War, there were reports that the Chinese government was using "brainwashing" techniques involving sensory deprivation for torture and mind control. Canadian psychologists and the Canadian government were eager to understand these brainwashing techniques, so the government began funding psychological research on sensory deprivation.

Among those embarking on this research were Marvin Zuckerman and his lab at McGill University in Montreal.[16] In the typical experiment in Zuckerman's lab, participants would spend hours in environments where they could hear or see very little. In some cases, people would sit alone in a dark, sound-dampened room with nothing to do. They could leave only to get a lunch of cold sandwiches or to use the bathroom.

Just as interest in mind control had inspired research in sensory deprivation, otherworldly concerns also helped shape the methods of this research. Zuckerman's lab adopted the Ganzfeld Procedure, a method of approximating sensory deprivation, to carry out some of the study. Wolfgang Metzger created this procedure in the 1930s, hoping it would release ESP abilities hampered by outside stimulation.[17] It casts subjects into a fuzzy nothingness, into what Metzger called unstructured sensations. Cut a ping-pong ball in half and tape the halves over your eyes while listening to static in headphones if you are curious about what it feels like.

Zuckerman's curiosity was piqued by how people reacted to the loss of sensation. For the first hour or so, all of the research subjects simply sat in the nothingness. But after that, things changed. Some sat quietly for hours upon hours. Others fidgeted, squirmed, and became bored and anxious, among other things.

Strangely, no existing psychological test could reliably predict how subjects would react to sensory deprivation. Zuckerman and his colleagues speculated that some people were high sensation-seekers and some were not. High sensation-seekers, they figured, needed high amounts of stimulation and were irritated by sensory deprivation. Meanwhile, low sensation-seekers weren't bothered by the lack of stimulation.

Yet even though Zuckerman's team theorized that sensory deprivation irritated high sensation-seekers, exactly this group of people signed up for the experiment in droves. These prospective subjects surprised the researchers not only because they were high sensation-seekers who should have found the study boring and frustrating, but also because they were non-conformists, a group the researchers would never have imagined to be interested in a tedious scientific experiment. These were the early 1960s, where combed slick, closely shorn hair was the norm for men – yet many guys with motorcycle jackets and long hair were eager to volunteer for the study. Why would an experiment in dullness bring the "hippies" out of the woodwork? The researchers were stumped.

Apparently, information had circulated to these eager volunteers: the sensory deprived experience had induced hallucinations for some of the early participants. The newcomers were there to seek the sensation of the hallucinations, not for scientific advancement or for financial compensation.

Zuckerman realized that sensation-seeking was not only a quest for external stimulation as they originally thought. It seemed as though high sensation-seekers wanted unique experiences, too. He asserted that sensation-seekers are sensitive to their experiences and choose stimulation that maximizes them. Sensation can come from emotions, physical activities, clothes, food, or even other people. Someone with a high sensation-seeking personality actively pursues experiences.

Because of this active pursuit of new experiences, sensation-seeking doesn't just describe reactions to a sensory deprivation experience. Sensation-seeking can reach into every aspect of life. It can affect your choice of activities, the way you interact with other people, the things you do for fun, the music you like, the jokes you make, and even the way you drive.

If you think of sensation-seeking as a continuum, high sensation-seekers are at one end. They are always seeking new experiences, even if (and in some cases because) they come with risks. Low sensation-seekers, on the other hand, may actively avoid new experiences. Most people, as you can imagine, fall somewhere in the middle, seeking out new experiences unless there's something to lose by doing so.

Zuckerman suggests that "sensation seeking is a personality trait defined by the search for experiences and feelings, that are varied, novel, complex and intense, and by the readiness to take physical, social, legal, and financial risks for the sake of such experiences."[18] Even though Zuckerman didn't learn anything about mind control with these experiments, he learned a lot about the various components of sensation-seeking that we continue to build on today.

Zuckerman created a sensation-seeking scale to assess how much of a sensation-seeker someone is overall and how they score in each of four subtypes of sensation-seeking (a concept I'll talk about more in a moment). Zuckerman's sensation-seeking scales have evolved over time from a general scale in the early years to the current, more complex version (known as Form V). I've put a copy of this in Appendix 2.[19]

It's worth noting that some people have taken issue with Zuckerman's research, his definition of high sensation-seeking, and the way his scale is structured in particular. For example, the quiz is in what is called a "forced choice format" – this means you must pick between one of two statements that best describes you ("I like 'wild' uninhibited parties" or "I prefer quiet parties with good conversation"). Researchers like Jeff Arnett have suggested this forced choice format doesn't allow for any shades of gray and that a Likert-type format where you judge each statement based on how well it describes you (i.e., "describes me very well," "describes me somewhat," etc.) would be more effective.[20] Arnett and others have created these Likert-type scales, and they work well enough, but the research says they don't seem to work any better (statistically) than Zuckerman's original scale. (You'll find Arnett's alternative scale in Appendix 3.)

A few people have commented that dated terms reflecting the idioms of the 1960s and 1970s were included in the original scale (hippies, queer, and others). This problem was rectified when the scale was updated in the 1980s to reflect the slang of the time. As we're now well into the 21st century, another update may be in order. The language doesn't necessarily impact the validity of the scale, however, even if it may impact the ease of understanding of those who take it.

Probably the most valid concern about Zuckerman's research centers on the confounding factors that are built into the scale depending on how it's used. For example, research has shown that high sensation-seekers have a higher tendency to try recreational drugs, like marijuana and cocaine. If you're trying to figure out whether or not high sensation-seeking and drug use correlate, you can't offer options like "I would not like to try any drug which might produce strange and dangerous effects on me" and "I would like to try drugs that produce hallucinations," because they ostensibly relate to both variables being tested (high sensation-seeking and drug use) and therefore explain away some or all of the potential correlation. I think this is probably an issue from the standpoint of statistical accuracy, although Zuckerman has rebutted this concern saying that validity in predicting engagement in certain behaviors (like sex, drugs, and others) has been unaffected even when the offending items are removed.

Rick Hoyle and his colleagues have revised Zuckerman's classic scale into a version called the Brief Sensation-Seeking Scale (BSSS).[21] The BSSS addresses some of the concerns of the classic scale but it still reveals scores on each of the four subscales. If you're interested in finding out where you lie on the continuum of sensation-seeking, you can take Zuckerman's classic scale in Appendix 2 or Hoyle's BSSS below. It's really best if you take the quiz before you read the rest of the book in order to gain the most insight into your own sensation-seeking qualities. Just note your scores for now, we'll interpret them in a moment.

The Brief Sensation-Seeking Scale

For each statement, describe yourself by picking a number from 1 to 5, and add up your answers as directed.

1	2	3	4	5
Not at all like me	Not like me	Unsure or both like and not like me	Like me	Very much like me

	My Score (1 to 5)
1. I would like to explore strange places.	
2. I would like to take off on a trip with no pre-planned routes or timetables.	
Add up your scores from statements 1 and 2 and write the total in the box on the right	ES Score
3. I get restless when I spend too much time alone.	
4. I prefer friends who are excitingly unpredictable.	
Add up your scores from statements 3 and 4 and write the total in the box on the right	BS Score
5. I like wild parties.	
6. I would love to have new and exciting experiences, even if they are illegal.	
Add up your scores from statements 5 and 6 and write the total in the box on the right	DIS Score
7. I would like to try bungee jumping.	
8. I would like to do frightening things.	
Add up your scores from statements 7 and 8 and write the total in the box on the right	TAS Score
Add ES + BS + DIS + TAS Total Sensation-Seeking Score	

Adapted from Hoyle, R. H., Stephenson, M. T., Palmgreen, P., Lorch, E. P., & Donohew, R. L. (2002). Reliability and validity of a brief measure of sensation seeking. *Personality & Individual Differences, 32*, 401–414. Copyright (2002), with permission from Elsevier.

Zuckerman's way of looking at sensation-seeking isn't bad when you weigh all of this against what other theories have to offer. What's more, it's a body of work that has withstood the test of time and scientific scrutiny. Zuckerman was one of the first scientists to attempt to identify and understand the high sensation-seeking personality, and his work remains not only the most valid, but also some of the most interesting in this area. I've used his framework to better understand the high sensation-seeker throughout this book, so let's explore that framework more closely and then you'll get a chance to interpret your scores.

The Components of Sensation-Seeking

Zuckerman recognized that the high sensation-seeking personality is complex. It is made up of four distinct components, each of which contributes to an individual's unique way of seeking or avoiding sensations:

- thrill- and adventure-seeking: the quest for risk and danger;
- experience-seeking: the love of variety in sensations of the mind and senses;
- disinhibition: the ability to be unrestrained; and
- boredom susceptibility: the dislike of repetition.

Let's review each one.

Thrill- and Adventure-Seeking

When you think of sensation-seeking, thrill- and adventure-seeking probably come to mind. This component of sensation-seeking emphasizes the enjoyment of at least moderately frightening activities. The secret in the sauce of thrill- and adventure-seeking is the potential for danger. Those with high thrill- and adventure-seeking personalities seek out physical activities that are exciting and risky. For some, the risk is not an essential part of sensation-seeking; it's just the price of admission for the novelty that many people with high sensation-seeking personalities crave.

Both Mike and Kirill fit this profile perfectly. They know the risks, but do things like skydiving and hang gliding anyway. Mike, in particular, scores off the chart for thrill- and adventure-seeking – a ten, the highest score possible. "The first time I went skydiving was in high school," he explained. "I was interested in possibly going, but never really took that first step, but then a friend of mine was turning 18 and he asked me if I wanted to go, and I said, 'Alright, I'll try it out.' Eventually I went with him and did my first tandem skydive. It was kind of cool cause it's a no control situation at first. When you're in free fall immediately followed by a serene sort of feeling . . . the closest thing to humans flying."

This is the status quo for thrill- and adventure-seekers. Risks may be ignored, tolerated, or minimized, and may even be considered to add to the excitement of the activity. In contrast, those who do not seek out thrill- and adventure-seeking may avoid activities that seem risky or dangerous.

Experience-Seeking

Even if you're not an extreme thrill- and adventure-seeker, there may be a component of sensation-seeking that applies to you. So while you may not like to skydive, you may still exhibit a sensation-seeking trait associated with people who enjoy new, complex, and intense sensations and experiences. It's called experience-seeking.

Experience-seeking is the quest for new experiences that challenge the mind and senses. Experience-seekers look for a variety of experiences that are unique, rather than dangerous. These experiences may affect sensation-seekers emotionally, intellectually, or interpersonally, through all five senses, including sight, sound, taste, touch, and even smell. Think of experience-seeking as internal sensation-seeking.

A few years ago, I was headed to Hong Kong for a week. It was my first time there so I decided to hit up the *New York Times* for some ideas of what to do. They have a feature in their travel section called "36 hours" where they give advice about things to do if you are only going to be in a location for a short time. They suggested a restaurant that was pretty difficult to book. It took my friends 3 months to get reservations. A few weeks before we left, I got an email confirming our reservation. It read, "We will delight you with a 6 course exotic meal." That sounded good. It went on, "and we'll present you with the menu of the meal at the end of the meal." What? The END of the meal? If you score high in experience-seeking, I bet you are thinking, "Sure, I'll try anything once . . . that sounds kind of fun." If you, like me, score low in experience-seeking, it's likely that you want to know exactly what things are BEFORE you put them into your mouth.

Remember Sophie from the start of the chapter? While she scores low in thrill- and adventure-seeking – she wouldn't go skydiving again – she scores high in experience-seeking. For Sophie, the satisfaction comes from pushing herself in order to conquer challenges. What drives her to bike and travel is challenging herself. Her challenges are really a thirst for new experiences and not a quest for danger.

"I did go skydiving once, I kind of enjoyed it, but it made me realize that I mostly seek thrills and a sense of satisfaction from the types of things I do like – adventures and activities. For me, my experience could be going for a walk in the morning and seeking the sunrise or something. I always want to challenge myself. I love traveling, I love to meet new people, I love new experiences. The challenge is the quest to experience the unknown and to see where it takes me."

Another aspect of experience-seeking involves being around people who stimulate these experiences because they are unpredictable and different. Experience-seekers are attracted to an anti-establishment personality type that rejects the norm. We'll discuss more about this in Chapter 5.

Disinhibition

Disinhibition involves our ability to be spontaneous. It includes searching for opportunities to lose inhibitions. People with strong disinhibition tendencies act regardless of potential consequences, while people with low disinhibition tendencies control their behavior more carefully and think through more of the consequences. They look before they leap. People high in disinhibition? They just leap. I know it sounds a little backwards, but the higher your disinhibition score, the more spontaneous you are likely to be. It might not be surprising that these people are more injury prone and more likely to participate in activities such as the World Naked Bike Ride or the Burning Man Festival.

Remember Sophie, the lover of challenges? Her scores for disinhibition are relatively high, and it shows. One of her favorite things to do is "leave home with my bike, passport, and credit card, and go where the wind takes me." That's a great plan for someone who scores high on disinhibition (meaning no plan). Her scores for disinhibition show in other ways too.

In a blog post Sophie describes how she ran 26.2 miles, baring her story and exposing her inner demons "in hot pants, a sports bra, and covered in paint."[22] Rather than running the 2016 London Marathon in typical sports gear, Sophie decided to use the opportunity to tell her story by painting her body with inspirational messages and even some of her fears (and a sprinkling of gems and glitter too). She had her mother snap a photo for Facebook. Her mother tried to talk her out of it, fearing she was making herself too vulnerable. Sophie's response? She did it anyway. She explained "There's no point in going half way in, to test the water to see what happens. It has to be all or nothing, and that's the way I live."

She's not alone. Lots of high sensation-seekers with high scores in disinhibition live life this way. When they try a new food for the first time, they don't take a nibble, they open their mouths and take the biggest bite possible and deal with the fallout later.

Boredom Susceptibility

The last component of sensation-seeking is boredom susceptibility, which boils down to one's ability to tolerate the absence of external stimuli. Those with high scores in boredom susceptibility dislike repetition and they get irritated when nothing is going on. People with high boredom susceptibility also tire easily of predictable or boring people, and they get restless when things are the same.

My scores for boredom susceptibility hover near the bottom. I almost never get bored. I remember waiting in line for the original iPhone, over a decade ago. At that time you couldn't order it in advance. You had to wait in line to be one of the first to get the device. So I did. I waited in line over 5 hours with not much to do (I couldn't fiddle with my phone – because I was waiting for one of the first smart phones). It was a situation similar to what Zuckerman's original sensory deprivation participants had to endure, except I had a Cinnabon across the way. How did I cope with the lack of entertainment? Easily. Even telling this story to my friends with a higher boredom susceptibility score seems to irritate them.

Interpreting Your Scores

The Brief Sensation-Seeking Survey (or BSSS) reveals five scores: one for each component of sensation-seeking and a total sensation-seeking score. Here is what the scores mean.One thing to keep in mind: These ranges are simple comparisons to the general population. Don't get too worried if your numbers are low or high. Think of the score as a way to understand the relative nature of sensation-seeking. There are no right or wrong scores or best way to be. The scores are just a clue to your individual pattern of seeking or avoiding sensations.

Let's take a look at the overall score first. Overall scores on the BSSS will range from 8–40.

- If you scored between 8 and 16, you are a low sensation-seeker. You are more likely a chill-seeker than a thrill-seeker.
- A score from 16–28 is average. An average sensation-seeker may enjoy some new things but doesn't want to get too stimulated.
- Anything over 28 means you are a high sensation-seeker.
- You are more likely to feel a need for new experiences or stimulation. You may also find yourself easily bored.

But the total score doesn't tell the whole story. Your score on each of the four components will explain the *type* of sensation-seeker you are.

Thrill- and Adventure-Seeking

- Thrill- and adventure-seeking is the desire to engage in activities involving some physical danger or risks.
- For thrill- and adventure-seeking, a score under 4 means you enjoy calmer activities.
- Between 4–7 means you like activities that get your blood pumping.
- Anything over 7 means you are drawn to the most exciting experiences.

Experience-Seeking

- Experience-Seeking is the quest for sensations of the mind and the senses.
- Experience-Seeking Scores under 4 mean you are more comfortable with familiar experiences.
- Between 4 and 7 means you like balance with experiences of the mind and of the senses.

- Over 7 – well that means you crave the most interesting experiences.

Disinhibition

- Disinhibition is your ability to be unrestrained.
- Lower scores, under 4, mean you like to have fun but have clear limits.
- Between 4–7 indicates you go wild at times, but you won't do just anything.
- Over 7? You may not stop to ask what people think before doing something that sounds fun.

Boredom Susceptibility

- Boredom susceptibility describes how easily you get bored and how irritated you become when bored.
- People who score under 4 can stay with the familiar for a long time.
- Between 4 and 7 means you may like traditions, but you need variation every now and then.
- Over 7? That means you may get easily bored and crave something new.

Overall Score Profile

Overall scores on the BSSS will range from 8 to 40:

- < 16: Low Sensation-Seeking
- 17–27: Medium Sensation-Seeking
- 28–40: High Sensation-Seeking.

For Each Subcategory

Scores for each of the components range from 2 to 10:

- < 4: Low
- 4–7: Medium
- 7–10: High.

Variations in Sensation-Seeking

As you can imagine, how you score in any given area is likely to predict the kind of behaviors in which you engage. So let's take a look at some real-world examples from people who took the sensation-seeking quiz on my website and wanted to share their stories (I've changed their names).

Name: Carl
Total Sensation-Seeking: 19/40
Thrill- and Adventure-Seeking: 5/10
Experience-Seeking: 4/10
Boredom Susceptibility: 5/10
Disinhibition: 5/10

I love roller coasters and stuff like that. I thought it might be fun to learn to fly a plane but when I start thinking about it, I get too scared to do it.

Carl is typical of a lot of people. With average scores for all four components of sensation-seeking, he'll seek out interesting or thrilling activities as long as they aren't too dangerous.

Name: Aaron
Total Sensation-Seeking: 19/40
Thrill- and Adventure-Seeking: 2/10
Experience-Seeking: 3/10
Boredom Susceptibility: 2/10
Disinhibition: 3/10

I kinda don't understand thrill-seekers. I'm the kind of person who finds something interesting in everyday stuff. I can listen to the same music over and over again and I love going to my favorite restaurants and ordering my favorite meal. They see me coming and get it ready.

While Aaron may be generally curious, I wouldn't wait for him to volunteer to do anything even remotely risky or dangerous. I would assume that he's generally curious but he wouldn't go out of his way for something unusual. On the contrary I would assume he might avoid things that are too unusual or dangerous. His low scores for boredom susceptibility explain why he can eat the same meal over and over again and enjoy it thoroughly.

Name: Rachel
Total Sensation-Seeking: 29/40
Thrill- and Adventure-Seeking: 3/10
Experience-Seeking: 8/10
Boredom Susceptibility: 8/10
Disinhibition: 9/10

I avoid all things that I feel are completely out of my own control. But I love doing random and spontaneous things where I'm in charge. Hence, I *hate* flying commercially and have a fear of flying.

While Rachel's total score is higher than average, her thrill- and adventure-seeking score is pretty low. With a thrill and adventure score of only 3 you aren't going to see Rachel doing dangerous things. But, like Sophie, you're likely to see her traveling or seeking out other new and intense experiences.

Name: Alex
Total Sensation-Seeking: 35/40
Thrill- and Adventure-Seeking: 10/10
Experience-Seeking: 9/10
Disinhibition: 8/10
Boredom Susceptibility: 8/10

My family has described me as suffering from 'pathological wanderlust' and always seeking adventure. When I was younger, they'd jokingly (sort of) try to take away my *National Geographic Adventurer and Traveler* magazines because they gave me 'bad ideas.' When I'm sick and in bed, I watch travel and adventure documentaries. I'm not very active on social media, because I get bored sitting at my computer.

It's not surprising that Alex might be described as having pathological wanderlust. All of her scores are quite high including thrill- and adventure-seeking, so she might take risks, her experience-seeking might show itself as wanting to travel, but she gets bored easily and is pretty disinhibited. I wouldn't be

(cont.)

surprised if she didn't get herself into trouble every now and then, which is probably why her family calls her wanderlust pathological.

While Rachel's and Alex's scores are both in the high range, you can expect the way in which they would exhibit sensation-seeking to be quite different. This is true across the board. People score differently in these areas and their scores have been shown over and over again to reliably predict what they are likely to do. Zuckerman's tool for judging this turns out to be a simple and elegant way to describe some of the nuances in sensation-seekers. It's no wonder so many researchers have used this model to discover the inner world of sensation-seekers. It's always nice when you hit on that combination as a scientist. Despite the individual nature of the high sensation-seeking personality, there are some tendencies we see in the population at large.

Who are the High Sensation-Seekers?

Scores on each of the components of sensation-seeking increase with age, to a point. It's higher in children and continues to increase until the late teens and then starts to decrease as you get older.[23] This might explain why people like Mike and Kirill and those starring in YouTube videos hurling themselves off houses and lighting themselves on fire, tend not to be in their fifties. They tend to be, more often than not, males in the prime of their testosterone producing years (more on that in the next chapter). There is one exception – boredom susceptibility. For most people, boredom susceptibility remains relatively stable throughout life.

In general males have higher scores than females on three of the four parts of sensation-seeking.[24] Men tend to score higher on thrill- and adventure-seeking, boredom susceptibility, and disinhibition. Women outscore men only in experience-seeking. The gap, however, is closing in thrill- and adventure-seeking. The last few years have seen a steady increase for the scores of women in thrill- and adventure-seeking. You can see this reflected in the number of women participating in extreme sports. While I don't believe the popularity of high sensation-seeking as a whole is the result of changes in society, I do think this part of the puzzle probably is. It seems as though we are seeing a cultural shift in thrill- and adventure-seeking in the United States for women. There were activities previously "off limits" or discouraged for women. That is not the case any longer and more women are following their thrill-seeking desires and impulses.

~

It's clear that there are people in our world who don't just crave high sensation-seeking environments, they thrive in them. Whether it's cliff diving for Mike, or running a 110 km from London to Brighton for Sophie, or hanging off buildings for Kirill, these are the situations in which high sensation-seekers shine. And there are, in fact, tremendous advantages for those who covet thrills. People with high sensation-seeking personalities often feel a sense of calm when skydiving and are likely to feel relaxed in other high-pressure situations which begs the question, "Why?"

2 BORN TO BE WILD

In the summer of 2014, I was invited to Twin Falls, Idaho, to speak to a group of mental health counselors. I was there to discuss the details of the new psychological diagnostic system. Having never been to Idaho, I expected to encounter super-nice people, incredible landscapes, and potatoes. Twin Falls had something much more thrilling in store for me.

As I greeted the workshop participants, I mentioned my interest in sensation-seeking. One person chimed in, "Oh, then you'll want to visit The Bridge." I was confused. How exhilarating could a bridge be? Over the course of my visit, one person after another mentioned The Bridge.

The Bridge turned out to be the I. B. Perrine Bridge, which is famous for being one of the only bridges in North America where you can BASE jump year-round without a permit. Imagine, jumping off of a bridge with a parachute whenever you'd like. People travel from all over the world to The Bridge to BASE jump.

BASE – or Building, Antenna, Span, and Earth – is an acronym created from the favorite jumping off points of BASE enthusiasts. BASE jumping is similar to skydiving, but a bit more dangerous. Instead of dropping out of a plane, you plummet from fixed objects such as a building, antenna, span (bridge), or the earth (cliff) with a parachute or other specialized equipment. Why is BASE jumping more dangerous than skydiving? Well, sky divers start their descent from an airplane at heights ranging from 3,000 to 15,000 feet. Since you are falling from such a high altitude you'll get about 60 seconds of life-choice-questioning free fall before you deploy your parachute and start to slow down. BASE jumping is a little different. Most BASE jumpers leap from altitudes of less than 2,000 feet. Because they are jumping from a lower height, BASE jumpers don't have as much falling time. Jumping from say, 500 feet will only give you around 5.6 seconds from the building to the ground, which means you have to make your choices pretty decisively and pretty quickly.

My curiosity about The Bridge and BASE jumping got the best of me. On my way back to the hotel, I leaned forward and asked my cab driver to "take me to The Bridge." I didn't have to explain further; she knew exactly what I meant. Ten minutes later, we were there. As we approached, I saw only what looked like any other span of highway, with four lanes of traffic zooming by. Not a big deal. We pulled over to a viewing area, and once I had a moment to take in the stately truss and arch structure spanning the Snake River Canyon, I was awestruck. The Bridge soars some 486 feet above the Snake River. Imagine standing at the top of a 45-story building and looking down – way down.

I hopped out of the cab to look around. In a small clearing nearby, I found four people adjusting what looked like large kites on the ground. "They're jumpers," the cab driver rasped in a cloud of cigarette smoke.

One of the jumpers was Nick, a 22-year-old business owner in Twin Falls. He earns his living making custom parachute containers and other air sports accessories, and he moved to Twin Falls to be closer to The Bridge.

Nick said he'd been a thrill-seeker ever since he was a kid. "I started flying paragliders when I was almost 16. So I've been flying for a while. I started skydiving as soon as I turned 18, because you have to be 18 to skydive," he said, as if he still resented it.

"As far as building gear and stuff, I've just always done that – everything from go-karts to sailboats. I used to build ramps to jump my bike and skateboard. Anything to get me into the air. It was just a natural progression when I started jumping out of airplanes and BASE jumping that I would build things for that too," Nick shared. Jumping bikes turned into jumping out of planes. Jumping out of planes turned into jumping from bridges. "It's kind of cliché. You hear it from everybody... jumping forces you to be in the now, like nothing else matters, and you're just focusing on what you're doing and enjoying it."

Psychologist Mihaly Csikszentmihalyi (pronounced chick-SENT-me-high) refers to the feeling Nick described as a flow state: being fully immersed, holding an energized focus, full involvement, and enjoyment of the moment. In a flow state, a person is hyper-focused and connected to what they are doing. Their emotions are channeled, and they feel joy in the present. Csikszentmihalyi suggests that, in this state, a person is "completely involved in an activity for its own sake. The ego falls away. Time flies. Every action, movement, and thought flows inevitably from the previous one, like playing jazz. Their whole being is involved and they're using their skills to the utmost."[1] Csikszentmihalyi calls it "optimal experience." Most others refer to it as being in "the zone."

Turns out that flow is good for us, too. Adults who spend more time in flow are happier overall,[2] and they tend to be more cheerful, satisfied, creative, and have higher self-esteem.[3] Flow gives us enjoyment, and makes it easier to deal with stress.[4] Flow also enhances learning,[5] which could come in handy for thrill-seeking activities. Being in a flow state helps us notice little specific patterns in our environment, so we know exactly when to push the X button for the jump in Mario Cart and precisely when to deploy our parachute when jumping from a bridge. Just like Nick says, flow forces us to be in the now and flow also changes us.

Being in the zone can be addictive. Nick agrees, "If I don't do it for a period of time, I get stressed out and antsy, and I need to fly off of something." Nick high-fived me and climbed over the railing as nonchalantly as if he were stepping over a puddle on the sidewalk. Then, with no hesitation, he jumped.

It actually wasn't so much of a jump as a release. He gave himself to the air and fell.

But things didn't go as planned. Although this was the first time I had seen someone plummet from a bridge, even I thought his parachute opened a little late. "I knew it as it was happening," he said later. "I bought the parachute for $25 off of a friend. It's an old skydiving reserve that's older than I am. It just took longer to open than I figured. It's taking longer to open, so you continue to accelerate and free fall. I'm watching the ground rush up and it's almost like an eerie, surreal feeling. In a way, you're scared, but in another way you are really calm and taking it in. You're like, 'This is what's happening and I'm along for the ride,' you know?"

I didn't.

As a low sensation-seeker, I would have been so overwhelmed – flooded by the sensations and deafened by my own screams of terror – that I hardly would have noticed the ground racing toward me faster than it should have. And that's when my fight or flight response would have kicked in.

The fight or flight response is a biological reaction, shared by all animals. It governs how we respond to dangerous or stressful situations. If you've ever watched a nature documentary, you know what the fight or flight response is. That antelope being chased by a lion out on the savannah – that's the fight or flight response in action.

When your brain identifies stress it prepares your body for vigorous activity so it can handle that stress. Some people call it the fight or flight response because two of the most obvious reactions to immediate dangers are to fight them or to run from them. But that doesn't always work. Sometimes there's no one to fight, and no place to run (or flee). In many cases people just freeze. Because of this, some people now call it the fight, flee, or freeze response. I bet you've experienced this yourself, perhaps during a traffic accident, taking an exam, falling out of a tree and breaking your arm as a kid, or BASE jumping. In all of these situations and more, the fight or flight response is activated.

When your stress response gears up, it activates certain physiological actions in your body to get you ready to fight the dangerous activity, flee from it, or freeze to avoid detection. It's designed to mobilize our body as much as possible to help us to survive. The stress response involves a family of chemicals called catecholamines. Our most abundant catecholamine is dopamine, which is commonly linked to positive emotions, and norepinephrine and epinephrine (also known as adrenaline), which are linked to energy and arousal. Catecholamines prepare you physically and emotionally, heightening your awareness and focusing your attention with a surge of energy.

The body's stress response also releases corticosteroids, which help organize systems in your body to handle stress. These chemicals are produced in the adrenal cortex, which sits on the kidneys like a beret. Cortisol, the famous stress hormone, is an A-list corticosteroid. When cortisol is dumped into the blood stream, it's transported all around the body, and acts in varied ways on different cells and organs as it prepares the body to fight, flee, or freeze. It slows down some systems that aren't critical during the stressful period, while revving up the parts of your body that are crucial to survival (like your muscles). Under the influence of cortisol, your muscles tense, blood pressure increases, heart rate quickens, feel-good endorphins flow, pupils dilate, attention focuses, and digestive functioning slows to a crawl (which is why you get "butterflies" in your stomach when you're nervous). What's more, you breathe more quickly to get oxygen moving around your body. At the same time, blood is diverted away from your digestive organs and the bladder relaxes. This might explain why we feel the need to go to the bathroom during stressful situations. Blood is diverted to the major muscle groups. You are ready to rumble – or run.

For a low sensation-seeker like me, even using my phone to record Nick's jump was enough to get my catecholamines and cortisol flowing. The arousal caused by the situation was immense. Cars were zooming by on my left. The river was hundreds of feet below on my right. Nick and the rest of the jumpers were adjusting what seemed like dozens of straps and hooks that I imagined would lead to terrible injury if adjusted improperly. I worried that a sudden updraft might launch my phone over the guardrail as I tried to record the scene, not to mention people were climbing over a guardrail to jump off a bridge (after insisting on high-fiving me). My heart jumped in my chest. My breathing was shallow. I was clearly rattled – and I was just watching.

But when Nick jumped off the bridge, he was flowing instead of freaking. He said he felt calm and hyperaware. For high sensation-seekers like Nick, the extra arousal didn't overwhelm him. It enhanced the experience. How is this possible? It can be explained by what some psychologists call optimal level theory.

The Optimal Level Theory

Early theories of sensation-seeking focused on the "optimal" or perfect level of arousal for a situation.[6] It's like the story of Goldilocks and the Three Bears. You remember the story. Three bears, Papa Bear, Mamma Bear and Baby Bear, go for a walk after they've made their breakfast of porridge. Goldilocks wanders in and tries their chairs, beds, and breakfast, complaining that some aren't quite to her liking while others are "just right." The optimal level is like the Goldilocks zone – not too much, not too little, but just right. Too much stimulation can be overwhelming. Too little stimulation doesn't even get our attention. When arousal levels are just right, we are at our best. However, that perfect level can be very different from person to person. Some like it hot, some like it cold, while others, like Baby Bear, prefer something in between.

Optimal level theory can be traced back to one of the founders of experimental psychology, Wilhelm Wundt. In 1895, Wundt noticed that for various levels of stimulation there was an optimal level that was considered pleasurable for the subjects in his experiment.[7] Below or above the optimal level was judged less pleasurable or even aversive.

Let's take peppers as an example. My friend Debra is the Mama Bear of spicy foods. Even a little peppery heat is too much for her. The optimal level for her would be hardly any spice at all. That's different from another friend, Jack, who's more of a Papa Bear when it comes to spicy. He'll chow down on fistfuls of jalapeno poppers – something that would send Debra scampering for ice water while fanning her tongue.

It may be that Debra's optimal level is so different from Jack's because they have different levels of ability to perceive sensory information. Some researchers suggest that the more sensitivity to sensory experiences you have, the lower the stimulation you need to reach that optimal level.[8] These researchers would say that Debra's taste receptors are so sensitive that she needs very little stimulation from the spice to reach her optimal level. Jack, on the other hand, might be less sensitive, so he would need more spice to reach his optimal level.

The perception of the heat of the peppers may be like other sensations or sensory experiences. Perhaps high sensation-seekers are less sensitive to stimulation than most. That means that they need more stimulation to reach their preferred level of arousal, just like Papa Bear, who liked his porridge hot. Perhaps Nick only appreciates the experience of jumping off the bridge, so the theory goes, because of his high sensation-seeking personality. No low sensation-seeking person could tolerate that much stimulation, much less seek it out. Someone who actively seeks unusually high levels of stimulation must have a much higher tolerance than someone who is more sensitive. This means that Nick and other high sensation-seeking individuals crave highly arousing situations to reach their optimal level.

Even though his parachute opened late, Nick landed safely in the Snake River near the bank. Then he trudged up from his landing point to join his other jumper friends. "I've always romanticized flying," he told me later. "From the time I was 5 years old, I just always wanted to fly. I can't really explain it, it's just part of my DNA. It's just something I need to do."

Nick might not be too far off. Studies suggest that there's a strong genetic influence to sensation-seeking.[9] What is it about his genetic makeup that might lead Nick to be a jumper and to see the ground rushing up to him in a very different way than I would?

The Genes of a High Sensation-Seeker

Chris and Jess (you'll hear more about them later in the book) just had a baby. As high sensation-seekers themselves, they are hoping their new baby will inherit their love of rock climbing. Should they load up their new baby registry with brightly colored plastic toy carabiners and ascenders? Can we predict if the new baby will also have a fondness for heights? Predicting behavior isn't easy. In fact, much of it is a mystery. Mostly we think of behavior as something conscious, as when we choose what to wear from our closet or where to go on vacation. But behavior can also be automatic or even instinctual, like fear of falling from high places. This mix of intentional and instinctual behaviors is impacted by our biological makeup as well as environmental influences. Researchers in the field of behavioral genetics study variation in behavior as it is affected by genes. They study which differences among people can be explained in terms of genetic and environmental components.

Packed inside nearly all cells is a coded message. This message is locked away in deoxyribonucleic acid, or DNA. There are four nucleotide bases that make up your DNA: adenine, cytosine, guanine, and thymine or A, C, G, and T. DNA contains the directions to make every substance and structure in your body. There are more than six feet of tightly bundled DNA in each of the body's cells. In essence, your DNA contains the unique list of ingredients and the specific recipe to make you *you.* Anyone who has watched a crime drama lately knows that DNA is unique to every person who has ever lived.

Every part of your body is produced from this instruction manual, from earwax to endoplasmic reticulum and from toes to tears. It's the sum of your hereditary information, like an instruction manual for assembling your body. Your genes are the letters, each offering up a part of the story of you. Put the letters together in the genome and you have the story – provided you have put the letters together in the right order. Genes also produce the instructions for how to use materials. For example, they tell your stomach when to produce stomach acid and how much. If sensation-seeking is genetic, then the body must produce something physical that leads to this behavior. It could well be that something in Nick's DNA leads to his high sensation-seeking personality. However, DNA doesn't make personalities; it makes proteins.

The question is, what does the body produce (or not produce) in high sensation-seekers? Are they actually programmed differently than the rest of us? If so, how?

If you are hoping that one day they'll discover the gene for sensation-seeking, I wouldn't hold your breath. Complex personality factors like sensation-seeking are most likely influenced by what could be hundreds of genes, each interacting with the environment in sometimes unpredictable ways. Most research into genetics doesn't tell us which specific genes are involved, rather they attempt to describe the relative influence of all genes and every environmental factor. And by the environment we don't just mean the physical world. The environment encompasses all influences other than inherited factors, including family, friends, home, work, even specific experiences from everyday life, such as what we eat, and things we are exposed to like diseases or toxins.

But what is the best way to determine if Chris and Jess should sign their baby up for summer rock climbing camp? How do we determine if Nick's genes or something in his environment are responsible for his love of leaping? Too bad he doesn't have a twin.

Twins, especially those who grew up separately, are perhaps the best way to determine the relative influence of biology and environment.[10] Twin studies are the gold standard of genetics research and are used to tease apart environmental and biological influences on traits and behaviors. Since identical twins share 100 percent of their genes, and twins raised together share 100 percent of their environment, examining twins in different families can be the ideal way to differentiate the relative strength of genes versus environmental contributions.

From twin studies you can determine what's called a heritability estimate. It's called a heritability estimate because it describes the influence of heredity and how much of the variation of a specific trait might be due to genes versus the environment. Heritability estimates range from 0 (meaning that genes have no influence) to 1.0 (meaning that genes determine everything).

The Minnesota Study of Twins Raised Apart or MISTRA is a longitudinal study of twin pairs who were separated in infancy, reared in different homes during their formative years, and united as adults.[11] These twins went through some 50 hours of medical and psychological assessments over a six-day time span at the University of Minnesota Psychology Department and medical school. As part of this extensive battery they were asked to take Zuckerman's Sensation-seeking Scale and the resulting heritability estimates were around .55.[12] Others have had similar numbers up to .58.[13] This means that genetics have a strong influence on sensation-seeking.

Back to Nick; when Nick jumped off the bridge, he was flowing instead of freaking. He felt calm and hyperaware. Let's think about why. His brain was bustling from all of the dopamine being released – making it a very pleasurable experience. In the meantime, he might have less norepinephrine flowing through his body, which means he felt energized but not out of control. Cortisol was doing its work to keep his muscles taut and ready for anything and his attention sharp, but he didn't experience this as anxiety because he was responding less than the rest of us would to the serotonin. Put simply, for high sensation-seekers like Nick, this extra arousal did not overwhelm him. It enhanced the experience.

Which leaves us with a very important question. If the fight or flight response is designed to help us survive, why did some of us evolve to respond to it differently? Jumping off a bridge is potentially life threatening – so are many things that high sensation-seekers do. It seems like high sensation-seeking would have self-selected out of the gene pool a long time ago with too many people falling to their doom jumping off cliffs. Here is the paradox: There may be an evolutionary benefit to it.

The Evolutionary Benefits of Sensation-Seeking

The central idea of Darwin's theory of evolution is pretty straightforward and also sometimes misunderstood. When you mention Darwin, people often blurt out "survival of the fittest," which makes it seem that evolution is some sort of biological cage match pitting specific biological features against each other which would result in the perfect version of every living thing existing in harmony with the environment. Not exactly. The general idea is that some biological features that an organism possesses might help the organism adapt to their current environment. For example, the ability to digest certain kinds of foods or even to react in a certain way to a dangerous situation. Also, mutations in genes occur all the time. Think about freckles, stork bites, or birthmarks. Those are often the cause of random gene mutations. Children aren't exact copies of their parents. Unless they are identical twins, each child has a unique combination of genes from both parents. Combinations of genes from parents to offspring result in the wonderful variations we see in all living things. Over time these variations might help or hinder survival. In psychology, the influence of Darwin led to an early school of thought called functionalism.

Back in the 1900s psychology was very different than it is today. A major way of thinking about behavioral and mental processes emerged called functionalism. Functionalism was concerned with the purposes of behavior. William James, an early functionalist, was inspired by Charles Darwin's work on finding meaning behind the behaviors and physical structures of animals. Darwin inspired biologists not only to describe but also to explain why animals have certain biological features such as long necks, or the ability to fly. The goal of functionalism was not simply to identify a behavior but also to describe how aspects of it increased the likelihood of survival. Functionalists were interested in how behavior acts to adapt to the environment. Modern functionalists or evolutionary psychologists try to explain the evolutionary benefit of our behaviors.

How does this relate to sensation-seeking? Many human actions contain some element of risk. Riskier actions can lead to great rewards or great peril. For example, being an experience-seeker might mean you are drawn to unusual foods you've never tried before. But choosing the wrong food can lead to disgust, or even death, if you've made a particularly unfortunate selection. For low sensation-seekers like me this risk would have been an incentive to stick with familiar, safe foods. Sensation-seekers may have chafed at a boring diet and sought new options, however, even if they seemed unlikely to yield much interest. As delicious as a pineapple is, on the outside it looks like a prickly pinecone. If I didn't know what was inside, I would likely not explore any further. Long ago, some curious high experience-seeker must have opened it and then enjoyed the reward of finding out that it was moist, sweet, and tart (and don't get me started on passion fruit).

Sensation-seeking may well have led to the kind of risk-taking that helped humans travel, adapt to, and survive in new environments. Hunters willing to take greater risks may have gained greater rewards. For example, bigger game means a risk of danger and a risk of not having any dinner, but if you get the animal, you can feed the tribe for weeks. No guts, no glory – they say.

While there may be advantages to this type of experience-seeking, there is also a down side. The manchineel tree is a pretty lime green tree that can grow up to 50 feet tall with tempting green fruit. Found in the Florida Everglades, Central America, and in the Caribbean, it bears a sweet fruit nicknamed "the apple of death" that blisters the mouth and closes the throat when eaten. Coming in contact with any part of the tree can be fatal, whether you breathe in its sawdust or get squirted with noxious sap.[14] If your novelty-seeking draws you to this fruit, it could very well be your last new experience.

Finding the right balance is difficult. Take too many risks and you may be dead before you have a chance to spread your genes by having children. Remain overly cautious and you might have a hard time getting what you need (including kids). Life requires a bit of risk-taking to be successful and evolution favored humans who took at least some risks, but not too many. The question is how is this influenced biologically and does it have something to do with what is good for us?

Evolutionary psychologists say that humans tend to be drawn toward things that are good for us and fear things that could be dangerous. This might explain why some people have what seems like an innate fear of spiders or snakes (they can be deadly), and many people crave particular foods that have nutrients which the person may lack (called specific hunger).

Scientists have even discovered some biological systems that back up this claim. Specialized systems in the brain have evolved to reward some behaviors with pleasure while associating others with pain.[15] It's like a biological traffic light. Potentially rewarding things get the green light and we want to approach them. Things that could be punishing or that won't bring a reward get the red light and we want to avoid them. This is called the neural motivation system, and it regulates how sensitive we are to rewards and punishments. What's more, it helps us move toward, or approach, things that are rewarding, and move away, or withdraw, from things that are not rewarding or are otherwise unpleasant. Our neural motivation system is divided into two independent systems: the green light is the Behavioral Approach System (BAS) and the red light is the Behavioral Inhibition System (BIS) – one for approach and one for inhibition.

The BAS is sensitive to reward and is activated when we sense something that could be rewarding. It controls approach and approach-type behaviors. Guess which chemicals are released to make those behaviors rewarding? Yep, the catecholamines, including dopamine. You'll recall that dopamine is a neurotransmitter associated with pleasure. Your body makes potentially rewarding activities feel good.

If the BAS is the green light of motivational structures, then the BIS is the red light. The BIS is sensitive to threat and non-reward. It is activated during punishment, boring activities, or negative events and is related to sensitivity to punishment. It ultimately results in avoiding these kinds of situations (also quite convenient). The neurotransmitters associated with the BIS include serotonin, which is associated with anxiety and mood, and norepinephrine, which is associated with energy. It is also linked with cortisol release. Are you starting to see a pattern here? These are the same chemicals we discussed before when talking about the fight or flight response.

When a situation occurs, your body responds with one of the two systems. Imagine you are at a restaurant and the server is describing some of the choices: "delicious shrimp cocktail!" You hate shrimp and are allergic to them. The Behavioral Inhibition System kicks in to red-light that option. You might even make a "yuck" face. "Seventeen-layer chocolate cake," she continues. Your favorite! Suddenly your Behavioral Activation System kicks in, dopamine is active and you are excited to order it.

If this is the case, then there are two other important biochemicals likely involved in this puzzle: the hormone testosterone and the enzyme monoamine oxidase (MAO). These affect your *reactivity* to the neurotransmitters dopamine, serotonin, and norepinephrine in your body.

Produced in both the testes and the ovaries, testosterone is the hormone responsible for the body hair, deeper voices, and often the odors, of adolescent boys. It fuels healthy libido, muscle mass, and energy levels. Getting the right amount in your system is important. When testosterone levels are too low, it can lead to fatigue, depression, and low sex drive.[16] These kinds of symptoms in men may partially explain the proliferation of Low T commercials on television and supplements that promise to boost testosterone levels.

Researchers have discovered a correlation between testosterone and sensation-seeking behavior. An April 1999 study in the *Journal of Behavioral Medicine* examined the medical records of over 4,000 men. Those with higher levels of testosterone were 24 percent more likely to report multiple injuries, 35 percent more likely to report having a sexually transmitted disease, and a stunning 151 percent more likely to smoke. These risk factors are clearly related to high sensation-seeking activities. And the higher the levels of testosterone, the more likely the research subjects were to engage in sensation-seeking behaviors.[17]

Some researchers connect the relatively higher levels of testosterone in men with their higher thrill- and adventure-seeking scores. What's more, several studies show a significant relationship between testosterone and aggression but interestingly only in people with low levels of cortisol (like high sensation-seekers).[18]

Higher levels of testosterone are associated with higher levels of sensation-seeking in both men and women.[19] Additionally, testosterone levels don't remain stable over time. They naturally decrease as we get older, and so do sensation-seeking scores. This might explain why many TV shows, movies, and amateur videos feature younger people doing thrill- and adventure-seeking stunts. Another chemical in your body that influences sensation-seeking is monoamine oxidase or MAO. MAO is an enzyme important in determining the levels of neurotransmitters that cross the tiny spaces between our brain cells called synapses. MAO is like a biochemical Pac Man, gobbling up stray neurotransmitters, the chemical messengers that help our neurons communicate. By gobbling up the errant neurotransmitters, MAO regulates their presence in our system. High levels of neurotransmitters increase neural activity, and low levels decrease it. High levels of MAO mean fewer neurotransmitters, which leads to low sensitivity. Conversely, low levels of MAO mean more neurotransmitters, which leads to high sensitivity.

There appears to be an inverse relationship between MAO and sensation-seeking.[20] When MAO levels are high, the neurotransmitters responsible for approach, such as dopamine, are eliminated quickly and our systems are less sensitive. This inverse relationship between MAO and neurotransmitters might be the reason that high sensation-seekers tend to have low MAO. This is consistent with other personality traits correlated with MAO. Low MAO is related to activity, sociability, and mania whereas the reverse is true for high MAO.

A few research studies have backed up this claim. One study looked at college students with very high and very low MAO levels.[21] The difference between low and high MAO subjects equated to time spent in social activities. People with low MAO smoked more, had higher rates of criminal offenses, more drug use, more alcohol use, and a higher rate of bipolar disorders.

A second study examined MAO levels in a group of people known for thrill- and adventure-seeking: bullfighters.[22] Bullfighting is an iconic tradition in several countries and traces its origin to at least 711 CE. If you don't know much about bullfighting except what you've seen in cartoons, you might think that it mostly involves waving a red cape in front of a bull. That's just one part of this much more complicated and rigorous sport. After an opening parade, there are three stages. In cape stage the matador is testing the strength of the bull with a series of taunts and passes with a cape. Then comes the picador stage where the bull is stabbed in the shoulders to anger it. And then finally the killing stage. Typical bullfights feature six bulls, three matadors and a crew of assistant bullfighters and can last a grueling four hours.

Jose Carraso and his colleagues randomly selected 16 professional bullfighters from a list of volunteers provided by the Society of Bullfighters of Madrid in Spain. They had the bullfighters take Zuckerman's Sensation-seeking Survey and took blood (the easy way) and measured their MAO activity. Not surprisingly the bullfighters scored higher in thrill- and adventure-seeking as compared to non-bullfighters on the survey. As expected, compared to other people of about the same age, bullfighters had lower levels of MAO activity and, of course, higher levels of sensation-seeking.

It seems to make sense, but on the surface, it doesn't explain why Nick would crave jumping off a bridge – something that's obviously dangerous. Wouldn't his inhibition system kick in to warn him that jumping off bridges is unsafe? Somehow it doesn't. Researchers are just now teasing apart how the arousal, approach, and inhibition systems might work differently in high sensation-seeking individuals. They have found some intriguing clues in non-humans, including rats.

High Sensation-Seeking Rats?

In the early 1990s, Françoise Dellu and his colleagues at the Scripps Research Institute in La Jolla, California, examined sensation-seeking in rats.[23] They wanted to know if rats, like humans, had different levels of sensation-seeking.

They used newly born Sprager Dawley rats - you would recognize them as the white ones with beady red eyes - commonly used for medical research because of their calmness, which makes them easy to handle. Researchers plopped them into a maze to see how they reacted in a stressful situation. For a human, it would be equivalent to being plucked out of your apartment and placed in a corn maze at random points during the day. The researchers were curious to see how the rats responded.

What they found were two very different kinds of rats. Some attacked the new environments with gusto, exploring every corner of the maze. Researchers called these high responders, or HR rats. Others seemed to be less active in these new environments and explored less. These were the low responders, or LR.

HR and LR rats seem to have different inborn "personalities." The HR rats did more than just attack the mazes more actively. They also explored more parts of the maze, ate their food more quickly, and, when given the opportunity through small tubes, delivered themselves stimulants more often. They were, in essence, high sensation-seeking baby rats.

What's more, the researchers discovered an important biological difference between the HR and LR rats. The stress hormones that were released served as a stimulant to engage in the stressful behavior. The researchers found that the HR rats' stress hormones were reinforcing to the HR rats. In other words, they seemed to seek out stressful situations. How could this be possible? Well, it seems that HR rats respond differently to stressful situations than their LR counterparts.

Like the rats, high sensation-seeking humans clearly respond to stress very differently than their low sensation-seeking friends. One explanation may be that high sensation-seekers respond differently to the neurochemicals associated with the stress response. Some researchers say that people with high sensation-seeking personalities tend to have a different sensitivity to the chemicals involved in the fight or flight response.[24] Mounting evidence suggests that high sensation-seekers react strongly to dopamine (pleasure), and weakly to serotonin (involved with anxiety) and norepinephrine (energy). Since dopamine is associated with pleasure and reward, serotonin with inhibition and anxiety, and norepinephrine with energy arousal, this would mean that high sensation-seekers may have a completely different response to stress than someone like me.

If we think through what we have learned so far about how Nick's body responds to the biochemicals that drive these choices, the picture starts to become clear. Dopamine is being released en masse when he BASE jumps, and he responds powerfully to it, so the action is more rewarding to him than it would be to most of us. Serotonin and epinephrine are also being released, but he responds weakly to these. That means the anxiety the rest of us would experience is diminished. In addition, it's entirely probable that Nick has high levels of MAO and testosterone, which we just learned further blunt his reactions to certain neurochemicals. The net result? In a situation where you would think his BIS would be flashing like a railroad crossing light, "Stop! Please stop!," the opposite is actually happening. His BAS is activated, and he's drawn toward the opportunity to fall off a bridge. Nick is experiencing his optimal level of sensation. Nick may actually feel less stress and more pleasure during risky and high sensation-seeking activities. This can mean the difference between a euphoric parachute jump and a terrifying one.

This is almost certainly due to the differences we have reviewed about how high sensation-seekers respond differently to stress, and the neurochemicals involved in the fight or flight response and the neural motivation system. But it could go further than that. Perhaps the physical sensation of falling is simply more excruciating to me than it would be to Nick. Maybe my very nervous system is different than Nick's, and I actually experience more physical sensation than Nick does – so much that falling off a bridge totally overwhelms my system setting off all of the terrified responses I've mentioned throughout this chapter.

Whatever the case, one thing we know for sure is that Nick doesn't seem overwhelmed in these incredible circumstances. It seems pretty clear his optimal level of sensation is *way* different than mine.

Biological Differences

Let's take a look at how these biological differences manifest themselves in each of the four subtypes of sensation-seeking.

Thrill- and Adventure-Seeking

Thrill- and adventure-seeking is the classic high sensation-seeking trait. The potential danger can be magnetic. For those with high sensation-seeking personalities, their increased release of dopamine during dangerous activities means they get more reward from thrill and adventure than do their lower sensation-seeking counterparts. And since high sensation-seekers have lower amounts of norepinephrine and thus lower arousal systems, they show lower amounts of cortisol in response to stress.[25] They actually feel less stress and more pleasure during risky and high sensation-producing activities. The same activities can push a low sensation-seeking person beyond their optimal level of arousal. For the low sensation-seeker, positive sensations shift to negative ones. This can be the difference between a euphoric jump and a stressful one. Nick is the classic example of this.

Experience-Seeking

The important component in experience-seeking is novelty rather than danger. For some, the uncertainty of what's to come brings anxiety and fear. It activates our behavioral inhibition system. For the high sensation-seeker with a weaker inhibition system, novelty isn't associated with anxiety. The weak inhibition system of high sensation-seekers means lower cortisol levels. Lower cortisol levels means a reduced stress response. The increased dopamine means those activities bring greater pleasure. Those who are predisposed to be experience-seekers are able to resist and minimize psychological and physical stress. They have a high threshold for aversive stimuli, and they are less irritable. This allows them to tolerate stress with new experiences better than lower sensation-seekers, which can be an advantage. In fact, a study of smokers found that those who were high experience-seeking tolerated smoking deprivation better than low experience-seekers.[26] When confronted with travel, new foods, and other non-dangerous options, the low levels of stress hormones and higher experience of pleasure make it easy for experience-seekers to try new things.

Boredom Susceptibility

People who are susceptible to boredom have a difficult time tolerating repetition and non-reward. For them, doing things once is enough, and if nothing is going on, they find it difficult to remain satisfied. For them a state of non-arousal is hard to tolerate. They seek out sensations to make up for it. This might explain Nick's frustration when he doesn't jump for a while. In contrast, low sensation-seekers experiencing the same low level of norepinephrine are more content and happy.

Disinhibition

Disinhibition is related to the inability to hold back. What typically holds us back from doing something is the potential for punishment, or even the anxiety around imagined punishment. I recently had lunch with a friend at a sandwich shop, just the two of us. We shared a booth. Between us was a sign that said "Booths Reserved for Parties of Three or More." Having low disinhibition made it difficult for me to sit there. It was as if the words on the sign got larger every time I looked at it. I imagined a confrontation with the server about my inability to read, or my disregard for parties of three or more, or my contribution to the demise of civilization. My inhibition system dumped massive quantities of cortisol in preparation for an event that never occurred. In contrast, as you might expect, low levels of cortisol are associated with disinhibition. With a low inhibition system and relatively low levels of stress hormones, there is less holding back the disinhibited sensation-seeker from the actions they wish to take.

Enough with Nature, What about Nurture?

Although genes carry a genetic code, people are not computers. Genetic codes do not unlock specific behaviors, but they do make some behaviors more likely to develop than others. Biology isn't your destiny; it is an influence. If up to 58 percent of sensation-seeking is biological, then this leaves a fair bit of room for variation based on nurture, or experience. After all, just because something runs in families doesn't necessarily mean that it's biological. Last names tend to run in families, but last names aren't genetic. Families also transmit traits environmentally too.

Experience does shape sensation-seeking. It appears that a great deal of stimulation, especially early stimulating activities during childhood, is related to high sensation-seeking in adulthood. Parents, for example, can have an impact on the development of sensation-seeking. Fearful parents may discourage exploration and encourage children to fear the unknown (like Marlin the Dad in *Finding Nemo*).[27] While other parents may encourage exploration.

This might help explain my relative lack of sensation-seeking. I remember terrible thunderstorms growing up. In hindsight, my mother may have been a teeny bit nervous about storms. I remember the drill: we would unplug the TV and other electronic devices, and we were forbidden to use the phone for fear that lightning would travel into the house and blow up anything that got in its way, including our brains, if we were unwise enough to have a phone conversation during a storm. Could my mother's nervousness about storms have affected my levels of sensation-seeking as an adult? After all, kids can learn to be frightened of the things their parents fear. In the same vein, overprotective parents can easily discourage kids from exploration, leading the children to fear new situations in which outcomes aren't predictable.

Highly stimulating family environments are associated with higher levels of sensation-seeking. Where do you find highly stimulating family environments? Highly stimulating childhood environments are common for only children and first-born kids (especially in larger families). These are both demographics where you will find higher rates of high sensation-seeking.[28] Jan Feji and her colleagues discovered that the larger the family, the higher the level of sensation-seeking.[29] Why? It could be that the larger the family the more stimulating (and chaotic) the environment and the less closely parents tend to monitor each kid. This may also stimulate independence. How is it that both only children and children that grow up in larger families have higher rates of sensation-seeking?

Jan Feji and Toon Taris suggested that it really isn't the environment, like birth order or family size that's important, but rather the parental styles that are created by these environments. The family environment has an effect on parenting behaviors which in turn have an effect on sensation-seeking.[30]

Religion seems to have an effect as well.[31] Those who say they are religious or spiritual and who engage in regular religious activities such as church or synagogue score lower in sensation-seeking. Researchers examined twins who grew up in different families and it seems that religious activities lowered sensation-seeking, especially the disinhibition component, and especially in males.

Clearly biological factors aren't the only things that influence sensation-seeking. The environment plays an important role too.

Above Genetics: What You Do Can Also Change your DNA

You can swap out many things in life – a salad for fries or even that strange tie you got from your sister for one of your birthdays. One thing you can't? Your DNA. You're pretty much stuck with the genes you got at conception, some 38 weeks before you were born. But there's a bit more to the story.

The Human Genome Project was a 2.7 billion dollar 15-year effort to map out all the base pairs that make up human DNA.[32] The Human Genome Project's goal was to transcribe those three billion base pairs in DNA and find out what makes a human. But what they discovered was that even after decoding all three billion letters, there was something missing. Turns out that DNA sequences are just the beginning when it comes to understanding genetic traits. Genetics simply determine what genes people have. But having a gene is just the first step. A gene has to be turned on to be helpful and does nothing if it's silent or turned off. It seems as though not every gene is activated and some genes can be switched off or silenced by the body. It's a process called methylation, where a small marker, just one carbon and three hydrogen atoms, is attached to a section of DNA and switches it off. These changes don't involve changes in the specific gene sequences. What does that mean? It means that the expression of that gene is changed without changing the genetic code.

What can cause methylation? Lots of things: food, sleep, exercise, and even our behaviors cause chemical modifications around the genes that turn the genes on or off over time.[33] Epigenetics is the director of this process. If you think about our DNA as the notes in a song, the notes may be the same, but how they are played depends on their expression. The process of silencing the gene is another layer on top of the genetic code, so it's called *epi* (from the Greek "above") genetics. The study of these kinds of changes in genes is called epigenetics.

Methylation is powerful. For example, when a honey bee is fed royal jelly, methylation causes the bee to grow larger, live longer, grow ovaries and become a queen bee.[34] Methylation patterns are also affected by behaviors. Let's imagine a mother rat (called a dam) has a mischief of pups. There are so many, some get lots of attention, like licks and grooming, others get ignored. Rats change their DNA methylation patterns based on how much attention they are given by their mothers when they are young. In pups whose mothers give them a lot of attention, a gene that helps modulate stress is activated. On the other hand, when a mother rat ignores or neglects the pups, the gene remains silenced. Because of this, pups of attentive mothers grow up to be less stressed and chill while the pups of neglectful mothers grow up anxious.[35] If you examined the genes themselves they would be identical. What's different? The gene's expression – which is determined by the behavior of the mother.

It's possible that early stimulating environments like those in high sensation-seeking families could create genetic expression that increases the possibility of sensation-seeking. Maybe Chris and Jess should start saving for rock climbing camp for their new baby after all.

~

My head was still buzzing as my taxi drove from The Bridge. It was hard to wrap my head around the fact that during a business trip I witnessed four people safely land from a 400-foot jump. It seemed like a random spontaneous act – but it wasn't.

The decision for Nick to move to Twin Falls and to jump that day was influenced by his biology, environment, and maybe a few other things as well. I wondered: for someone who craved such exciting experiences, surely this need for the buzz must affect other areas of his life as well. I asked him if it influenced his relationship with his friends. "You mean my non-jumper friends?" he asked. "I don't have any."

Then it hit me. When high sensation-seeking reaches a certain point – driven by biology and environmental circumstances – it's not just something you do on the weekend. It's woven into the very fabric of your personality. Born this way or not, it's who you are. This made me curious: What does the daily life of a high sensation-seeker look like? What about relationships? Is work impacted? I realized there was something valuable to learn from looking at the ways in which high sensation-seeking manifests itself on a day-to-day basis.

3 FASTER, HOTTER, LOUDER: THE EVERYDAY LIFE OF A HIGH SENSATION-SEEKER

In *Alice in Wonderland*, Alice follows a white rabbit down a rabbit hole and then her strange, trippy journey begins. The white rabbit represents an idea or concept which could lead to discovery. It actually pops up in many places including the *Matrix* films, *Star Trek*, *Jurassic Park*, and Stephen King novels. It's also the pen name of a 24-year-old high sensation-seeking adventure blogger who views herself as a white rabbit open to new and different ways to do things.[1]

"The white rabbit doesn't tell you what to do, it shows you," she explained. And that's exactly what she has set out to do. Her plan? Travel the world for free.

"It's an idea that started growing when I started hosting on CouchSurfing," she said.

CouchSurfing is a website that connects people who need a place to stay with people who have space in their homes. The idea is pretty simple. You offer people your couch to sleep on when they come to visit your city. The website makes money through fees you pay to become a member. However, once you are verified you don't pay any money to Couch Surf. In 2018 there were 15 million members registered on CouchSurfing and 400,000 hosts.[2] The White Rabbit was one of them. Around that same time she got the itch to travel. But travel can be expensive. So she decided to use her new found love for CouchSurfing to travel the world.

"I'm going to travel 300 days with 0 money. I'll follow the sun. I don't want to have money coming in or going out," she announced proudly.

It's easy to think of all the obvious ways that staying in unfamiliar cities on strangers' sofas for an entire year might not work out well, but she was unconcerned. My first thought was to say "be careful," but no one is careless on purpose – are they? The words jumped out before I had a chance to pull them back.

"Definitely the first thing everybody always says to me is, 'Be careful,'" she sighed. "As if I don't know I have to be careful. As if I'm some sort of naïve young girl who's just going to pack her bags and go and hope everybody's going to be happy and friendly. I know there's a lot of danger with it, but that's the thing I love the most."

Where is she looking forward to going? Buenos Aires for sure. And then there's Ubatuba, (between Rio de Janeiro and Sao Paulo).

"Apparently that's a little village somewhere in the rain forest, and a guy, he texted me on CouchSurfing and said, 'Hey, I saw your project. Maybe you want to come here and explore my point of view of the local life here, because it's really wonderful.' I'm like, 'Hell, yeah.'"

I asked her about Disney World. She frowned, "Not Disney World. No, definitely not."

~

One of the last things I hope for when I'm traveling is that there will be danger. But it's no surprise that the White Rabbit, like other people with high sensation-seeking personalities, who thrive in situations that others might find scary or overstimulating, seek out traveling experiences that include the thrill of the unknown. As this chapter illustrates, high sensation-seeking personalities influence a person's everyday habits in all sorts of ways, including the way one thinks, eats, spends leisure time, socializes, and yes, travels – and these habits come with both benefits and pitfalls.

Funny You Should Ask

I'm about to do something every comedian, even the youngest knock-knock joke novice will tell you never to do – explain a joke. But this isn't just any joke, this joke has been calculated to be the "Funniest Joke in the World." Not necessarily a joke that YOU will find funny, but through testing this joke has been more likely to be found funny than any other joke, worldwide. Essentially this joke won a scientifically created ultimate joke tournament.

In 2002, Richard Wiseman and his team embarked on a quest to find the world's funniest joke.[3] They created a website for people to submit jokes and a system for people to rate the jokes they read. The researchers were curious about what kinds of jokes people of different demographics, backgrounds, and countries would find funny. Over the 12 months of the project they received over 40,000 jokes and millions of ratings for them. This joke was the funniest:

> A couple of New Jersey hunters are out in the woods when one of them falls to the ground. He doesn't seem to be breathing, his eyes are rolled back in his head. The other guy whips out his cell phone and calls the emergency services. He gasps to the operator: "My friend is dead! What can I do?" The operator, in a calm soothing voice says: "Just take it easy. I can help. First, let's make sure he's dead." There is a silence, then a shot is heard. The guy's voice comes back on the line. He says: "OK, now what?"

Why is the joke funny? Like most jokes, it contains two essential components: an introduction of incongruity and the resolution of the incongruity. Let me explain.

Some psychologists believe that the earliest signs of humor can be seen in childhood.[4] Something threatening is suddenly reinterpreted as play. Think about how we get a baby to laugh. We attack them with wiggly fingers, we zoom our faces close to them, we even sometimes pretend we are going to munch on their bellies. Combine that with the fact that adults are some 40 times larger than infants and it's the equivalent of having an elephant appear out of nowhere, stampede toward you flailing its trunk, and then tweaking you on the nose. And what IF an elephant appeared out of nowhere, stampeded toward you flailing its trunk, and then just tweaked you on the nose. It might be funny (eventually). Why? Sigmund Freud might have speculated it has something to do with tension and then reduction of tension.[5] Freud suggested that our Id, the instinctual part of our personality, gets a kick out of anything that reduces tension. It really doesn't matter what it is; when tension is reduced, our Ids can't get enough. It could be the change in tone from a musical bridge to the chorus, the change in speed from going up a roller coaster to speeding down the hill or the sudden unexpected twist in a joke. Anything that reduces tension is fun for your Id.

Most jokes rely on introducing an inconsistency to produce the tension and then resolving the inconsistency to reduce the tension. The hunter joke is funny because "let's make sure he's really dead" could mean check to see if your friend is okay or let's kill him. "Let's kill him" is the tweak on the nose because it's unexpectedly absurd.

Contemporary humor researchers call this "incongruency" and "resolution."[6] Incongruency and resolution are the important elements in all types of humor. Sometimes the incongruity can be resolved completely (called incongruity-resolution humor) but in nonsense humor there is a surprising or incongruous twist and the punchline either provides no resolution at all or might even present a new or absurd incongruity. Like this joke, from Jack Handy: "If you ever fall off the Sears Tower, just go real limp, because maybe you'll look like a dummy and people will try to catch you because, hey, free dummy."[7]

The distinction between various types of jokes and humor depends, in part, on the type of inconsistency and the way the inconsistency is resolved. It depends on how we close the gap of the inconsistency. How does this link with sensation-seeking? Psychological researchers back in the 1980s used sensation-seeking to predict the kinds of jokes that sensation-seekers are likely to enjoy.[8] They created a humor test consisting of 50 jokes and cartoons which were rated on scales for funniness and aversiveness (how annoying or embarrassing jokes were was also rated). The kind of humor was also rated (incongruity-resolution, nonsense, or sex). A total of 448 subjects were involved. People who scored high in boredom susceptibility and experience-seeking didn't seem to like humor in which the punchline was easy to predict. They were more drawn to nonsense humor, which is obviously illogical. It involves non sequiturs or bizarre juxtapositions that often come out of nowhere. The more absurd the joke and the more unpredictable it is, the more these types of sensation-seekers tended to like the humor. Why? Sensation-seekers prefer this kind of humor because they can't see the punchline coming. Those who get bored easily can predict what the joke is going to be, they finish it in their head and move on. The tension never gets built up because the inconsistency is instantly resolved. They resolve the tension well before the punchline is delivered and the joke just isn't funny to them.

The study found that sensation-seeking not only influences the type of humor but also the content of the humor. This plays out in their jokes too. High sensation-seekers who score high in disinhibition aren't as offended by sexual overtones in humor and tend to find these types of jokes funnier than those who might score higher in disinhibition.[9]

If you aren't a sensation-seeker you may find nonsense humor silly because you can't find the logic in it. You might find sexual overtones in humor off-putting. This could lead to awkward situations when sensation-seekers are telling sexually provocative or nonsense jokes. Additionally, sensation-seekers may not always be aware when they have gone too far (we'll talk more about this in Chapter 7).

Vicarious Experiences

Music, art, television, or movies can involve exciting conflict, risk, and chaos that are particularly attractive for those with high sensation-seeking personalities. Even though these vicarious experiences are not thrilling in the sense of being risky or requiring intense focus, they attract high sensation-seekers because they are interesting, or "arousing," as psychologists say, touching on the experience-seeking aspect of sensation-seeking.

The more unusual, complex, or surprising something is, the more likely it is to have a higher arousal potential. Imagine someone walking through the mall dressed as Santa in March. This would have a high arousal potential. But the same thing in December would have almost no effect. It's the unusualness – the surprise – that causes the arousal.

But repetition of an arousing stimulus reduces arousal through habituation. Put simply, the more you're exposed to something, the less arousing it becomes. The first time I saw the movie *Final Destination*, in which the main character thinks he and his friends have cheated death, but death hunts them down and kills them in ever more gruesome and unexpected ways, it was so frightening that I didn't sleep for a week. But if I watched the same movie 50 times (please don't make me), the 50th time would be much less terrifying than the first. That's the power of habituation: it drains the arousal potential.

High sensation-seekers seem to habituate more quickly than low sensation-seekers to arousing stimuli. To test this, Patrick Litle measured the brain activity of students watching a 20-minute segment of a horror movie.[10] While nearly everyone reacted to the clip at the beginning, high sensation-seekers' reactions plummeted near the middle. The clip simply didn't have the same impact over time as it would for low sensation-seekers. This might be even more true for people with higher boredom susceptibility scores.

Because media becomes tedious for them more quickly, high sensation-seekers are drawn to a wider variety of arousing vicarious experiences.[11] Though they tend to prefer live events like concerts and nightclubs, they also find stimulating sensations in movies, art, music, and video games. They watch TV less than low sensation-seekers, but when they do, they change channels more often and the higher the sensation-seeking score the more likely that they are drawn to graphic horror scenes and sports with high levels of contact.[12]

Jazz, Classical or Muzak?

In an unpublished follow-up to the classic Zuckerman sensory deprivation study, Zuckerman and Hopkins gave the subjects in a sensory deprivation experiment an option to listen to music in their rooms. They had three choices: classical, jazz, or Muzak (bland, non-offensive elevator renditions of popular songs). It's probably not much of a surprise to find that the high sensation-seekers chose jazz or classical music and the low sensation-seekers were more likely to choose Muzak.[13] Why? Many low sensation-seekers pick music for its ability to calm and soothe and high sensation-seekers choose music for stimulation and arousal.[14] The high sensation-seeker's penchant for complexity extends beyond background sounds.

How Complex is Too Complex?

Researchers were curious to discover if high sensation-seekers' love of complexity would carry over into the visual realm. They employed a test called the Welsh Figure Preference Test which consists of 400 black and white figures.[15] Respondents are asked to indicate whether they like or dislike each one. It is a pretty quick test, taking about 20 minutes at most to rate all 400 figures. What the researchers discovered were radical differences in design preferences for low and high sensation-seekers. Low sensation-seekers tend to like figures that are symmetrical and simple and high sensation-seekers like complex, asymmetrical figures that are "sketchy and shady." The Welsh Figure Preference Test so consistently predicts high and low sensation-seeking that it could almost be used as a test for sensation-seeking.[16] Unfortunately, these preferences for complex shapes doesn't always predict preferences for art, painting, and design. Osborne and Farley, for example, had groups of art and educational psychology students sort cards featuring reproductions of famous paintings by how much they liked the paintings. Preference for highly complex painting was not related to sensation-seeking for art.[17]

Multitasking and Sensation-Seeking

As a psychology professor, I try not to dwell on what my students are doing during class, lest they think I'm psychoanalyzing them. Yet, despite my desire to ignore their behavior, I'm fascinated by what they're doing when I occasionally wander among them while I'm lecturing. Many look as if they are suppressing terror – as if I've emerged from a horror film, like the creepy zombie girl in *The Ring* – and I can see why. Some are innocently taking notes in a notebook. Others, however, are clearly not focused only on the lecture: their desks hold a notebook, a chemistry textbook, a computer with five or six windows open, and their phone for texting. "I'm multitasking," one student explained with an embarrassed look when I caught him amid a Netflix binge.

Many people call it multitasking, but I've always referred to it as "multi-slacking." Despite the fact that so many feel compelled to do several things at once, we aren't very good at it. Computers can expertly multitask – even my phone is capable of 600 billion operations per second. But people can only multi-*slack*, doing several things poorly, rather than doing just one thing well. Research suggests that sensation-seeking might be involved in the desire to multitask, just as it is in a number of other cognitive habits.

Researchers have found that when people engage in multitasking, it is often because they have difficulty blocking out distractions.[18] However, this isn't the only reason sensation-seekers multitask. Sensation-seekers actually have a strong capacity to focus their attention on a stimulus or task, and their attention is not diminished when they are bombarded with noise or beset by worries. High sensation-seekers can pivot rapidly and easily between tasks, which can be difficult for the low sensation-seeker like me.[19] This might explain why I am so distracted by my students' distractions. But distractions actually help some sensation-seekers perform better because the distractions prevent boredom. Boredom can interfere with a high sensation-seeker's ability to complete work just as much as distractions can prevent a low sensation-seeker from getting anything done.[20]

Nevertheless, multitasking can be dangerous. Because of some sensation-seekers' high susceptibility to boredom, multitasking may sound like a harmless component of sensation-seeking. Those who multitask the most are actually the least effective at it, while also feeling the most confident in their ability to do many things at once.[21] These self-assured, yet ineffective, multitaskers are also the ones who show high levels of sensation-seeking. For example, those with the highest self-reported cell phone usage while driving were multitasking sensation-seekers. A predilection for distractions paired with an overestimation of one's actual capabilities can result in dangerous situations on the road.

In their study of sensation-seekers, Joonbum Lee and his team had people tune their radios while in a driving simulator. They discovered that the higher the level of sensation-seeking, the more likely the participants were to tune the radio and glance away from the road. What's more, high sensation-seekers are more likely to drive faster (more on that in Chapter 7) and to glance away from the road when they are driving.[22]

Where the Wind Takes You: How High Sensation-Seekers Travel the World

According to a recent study, Americans spend nearly a year of their lives day dreaming about being on vacation.[23] It's no wonder that travel and tourism is one of the world's largest industries with a 2.3 billion dollar global economic impact. If you throw in accommodations, transportation, entertainment, and attractions, that number balloons to 76 trillion according to the World Travel and Tourism Council.[24] Why do people do it? All sorts of reasons. Travel is a common passion for many people, it helps them refresh their minds, some travel to visit a friend or a family member, or maybe it's the beauty of the location. The top reasons people are inspired to travel? Some 58 percent say they needed a break from their everyday life, 55 percent wanted to visit a friend or family member, 38 percent because of a life event, like a birthday party, anniversary or a wedding.[25]

Let's say you have a week off, and you can plan a vacation to anywhere you'd like to go. Where would YOU go? What would you do? Some people visit the same location year after year. Others might like to go to a city or a place they've never been. If you are in a place that's new, would you rather go on an organized tour and see the sights? Or wander around and explore all on your own?

It should come as no surprise that sensation-seeking might be able to predict what you might enjoy. It also should be no surprise that high sensation-seekers tend to prefer travel experiences that immerse them in a culture and experiences.

Remember the White Rabbit?

"It makes me feel alive," she explained. "It makes me feel like I'm actually using my brain. I'm actually thinking about, 'Okay, so this kind of situation, how am I feeling about that? What are the options of the things that can happen, and working with that?' It's amazing. It's an amazing thing to do."

For many people, when they feel fear, that's a signal to them that they shouldn't do something. For the White Rabbit it is the opposite. She thinks, "I should do it."

"Fear is something that I basically never have except when it comes to height. People ask me, 'Aren't you afraid, like to go travel?' I'm like, 'No. No.'"

My idea of a vacation is to find a pretty place and sit there, but high sensation-seekers would cringe at the idea of a do-nothing holiday.

In 1995, Helen Gilchrist and her colleagues at the Resource and Service Development Centre in Leeds, compared the sensation-seeking scores of a group of UK citizens who recently returned from an adventure holiday in Africa to a control group of a similar age and income. Their results won't surprise you. Highly significant differences were found between the adventure travelers and everyday people, with the adventure travelers having much higher scores in sensation-seeking, particularly the thrill- and adventure-seeking scale.[26]

This is certainly the case for Anne, a 43-year-old woman who recently packed up for an extended trip. "I was at a concert and

I just decided I wanted to do something really cool ... the music inspired me. I gave up my apartment and stored my stuff with friends. I had been reading about the South Pacific Islands and how they do the traditional navigation, you know by the stars. I was totally fascinated by it so I went out and I booked a ticket on a plane. I decided I'd be gone for a year. I got to Samoa and I started talking to some people who were about to set sail and asked if I could come along. It was pretty cool, it's like something you might daydream about, but I did it."

Why wouldn't most of us pack up for an adventure holiday to Samoa or a year of CouchSurfing around the world? For me it's because of the fear that it could go terribly wrong, you know, *Gilligan's Island*- or *Lost*-style wrong. That's not how high sensation-seekers see it. They tend to have positive expectations about their interactions with the world. Robert Franken and his team from the University of Calgary surveyed both high and low sensation-seekers and discovered a negative correlation between sensation-seeking and the tendency to see the world as threatening.[27] That means the higher the sensation-seeking score, the less likely one was to see the world and situations as potentially dangerous. Anne is no exception, like the time she encountered a snake on a recent trip.

"Last year I was traveling in Joshua Tree National Park and there was a big snake crossing the road, so I pulled my car over, stopped all the traffic behind me so they wouldn't run over the snake. Once the snake was at the side of the road I ran over and grabbed my camera, and I took pictures of it. The snake started striking at me. I thought, 'Oh my God, I've never seen a striking snake.' I'm getting all these good shots of it. Then someone pulled over and rolled down their window and said 'You're really brave!' That moment I realized I was actually in danger."

When I asked her what compels her to throw herself into unfamiliar and potentially dangerous situations she seemed puzzled by the question. "I trust myself," she said matter of factly. "No matter what situation I get myself into, I always find a way out."

Anne's examples are extreme but it's not that unusual for high sensation-seekers to travel spontaneously and immerse themselves in the cultures they visit. In 2004, Abraham Pizam and his team conducted a survey of over 1,400 people in 11 different countries and found that high sensation-seekers were pretty different in both their travel behavior and preferred activities as compared to low sensation-seekers.[28]

The researchers noted, "High sensation-seekers preferred active, spontaneous, fast-paced and less comfortable vacations. On the other hand, low sensation-seekers had a distinct preference for low-energy activities. They preferred passive, well planned in advance, slow paced and comfortable vacations" (guilty as charged). In fact, dozens of studies confirm this relationship.

Authentic Experiences

In 2006, Andrew Lepp and Heather Gibson at Kent State University surveyed over 200 travelers about their travel preferences. They found that "the higher the sensation-seeking score the more likely they were to have traveled to regions of the world rated as riskier." What's more, sensation-seekers,

> preferred more independent styles of travel and more novel activity than those lower in the sensation-seeking trait. High sensation-seekers travel freely without a well-defined itinerary, get off the beaten path, meet local people, engage in a host countries' culture and forgo comfort in favor of a more authentic experience. On the other hand, low sensation-seekers are likely to travel with packaged tours, pre-plan much of their trip, visit the famous sites, and maintain a barrier between themselves and the host culture and insist on familiar comforts.[29]

This goes along with the idea that high sensation-seekers aren't out for risk in and of itself. It's like the famous criminal Willie Sutton. He was once asked why he robbed banks. His answer? "Because that's where the money is." Why go to risky parts of the world and dive head first into an unfamiliar culture? Because that's where the unique experiences are. It makes perfect sense to forgo luxury for an authentic experience if it gets your dopamine pumping and your behavior inhibition system doesn't seem to mind.

Victor would agree. Victor is a 26 year old who's getting a PhD in Electrical Engineering focusing on artificial intelligence and robotics. He loathes dull vacations. In fact, he loves getting lost, because it's a puzzle to figure out. When I asked him what it was like hearing of a new adventure he said he actually craves collecting new experiences for what he called the museum of his mind:

"I'm sure anybody who's ever collected Pokemon cards or baseball cards or records or anything like that, when they come out with a new version of it, there's that initial perk of excitement, when you're like, 'Oh, interesting.' Then you think about how it fits into the rest of your collection. Every time something new comes up I think, 'Oh, wouldn't that be fantastic to add to that collection. Wouldn't that be a great other thing to go in that little museum I've got in my mind.' That's the experience that I feel when something comes up. Then after that it's immediately a sense of possessive desire to do it. Every time I learn about something new, I want to go out and try it."

There's another adventure that Victor loves. Food.

"I like trying all kinds of new stuff, despite the fact that I actually happen to have a minor gastric disorder. It can make me very sick, and certain things do. For example, my boss happened to bring in this exotic fruit. I'd never seen it before, never heard of it. I tried it, and it did make me extremely sick. I actually ended up having to go home that day. Then two weeks later, he brought in some other curio and it went in my mouth just as fast as that. Then he said, 'Shouldn't you think a little bit more carefully about that?' I was like, 'Nah, not really.' Then I tried it."

Some people call it food tourism.

Adventure Eating: Salivary Sensation-Seeking

Doing weird obstacle course races, scaling cliffs, jumping out of planes in a squirrel suit and flying around – these are the kinds of activities we typically associate with the high sensation-seeking personality. But eating? Do high sensation-seekers actually eat differently?

Some do. There are people who have a long-standing habit of trying new food. These are the foodies, the gourmands, the gastronauts. I'm not one of them. Think Anthony Bourdain. In case you haven't heard of him, Bourdain was a professional chef who wrote several best-selling books and then became a television sensation on shows like *No Reservations* and *Parts Unknown.*[30] The premise of all of these programs? Bourdain traveled around the world, encountering different cultures and investigating them, in large part, through their food traditions. As you might imagine, this led him to try some *weird* stuff – from the ortolan, a teeny bird that is eaten whole and extremely hot (once considered a delicacy in French cooking, eating ortolan is now illegal in the United States), to hakarl, the beloved (and, according to Bourdain, utterly disgusting) fermented shark of Iceland.

I'm not like that. I am pretty much the opposite. Plop me in most any restaurant and my choices are pretty predictable. I've even resorted, at times, to having my friends pick something for me. "You won't like that," they'll tell me if I consider ordering something unusual off the menu. This isn't new. I was a very picky eater as a kid. Peanut butter and jelly sandwiches were a staple in my lunchbox – so much so that my mom went on strike by third grade and I had to make my own lunches. Out of what I can only imagine as frustration, my family would drop me off at home after church on Sunday afternoons while they went out to eat. I imagined they saw this as a punishment, but for me making my own PB&J – for dinner no less! – was the ultimate luxury. I scored low on boredom susceptibility, so variety was a spice I used sparingly.

Fast-forward some 30 years, and I've become a teeny bit more adventurous about food. A few years ago I did an interview with Kate Sweeney, a local NPR reporter, about people who are fearless about food. She began by reading me what sounded like a list for an exotic party: "Black coffee, brambleberry crisp, goat cheese with red cherries, pad thai." It wasn't. These were the ice cream flavors we'd be trying that day. The shop she took me to is known for unusual and adventurous flavor combinations – not your typical 31 flavors, but ones that challenge traditional ice cream flavors. As a slightly more adventurous adult, trying a new flavor is a fun experience for me, but it's nothing like the flavors some of the high sensation-seeking foodies I've met will try. For them, food is an adventure in their mouths. They talk incessantly about food, chase down food trucks, and photograph and blog about their experiences. Some love food for the taste itself, or for the fads, or how it brings people together. They seek sensations in bowls of chicken hearts, goat brains, and pig blood stew, not because these foods are part of their cultural norms, but because they're there.

Munir, a former student of mine, reached out to me when he heard I was researching fearless foodies. He couldn't wait to tell me about some of the things he'd tried. "I love frog legs and gator," he started. "I don't know if you've ever had baby octopus – "

I hadn't.

" – but it's not cut up like squid or calamari. This baby octopus looks like they literally just pulled it out of the fish tank and threw it on the pan; then it's on your dish. It's like a baby octopus. Like it looks like a BABY OCTOPUS." He was already so excited he was nearly ranting.

"How does it feel when you're eating it?" I asked him. I was intrigued, but part of me didn't really want to know.

"So it's very small, like maybe the size of a quarter. It's chewy, very chewy and the head of it is almost like a chocolate Ferrero Rocher candy. But it's got this kind of rubbery skin, then all of sudden there's this gooeyness, and then it's like . . . I don't know what's in the center. I don't think octopus have brains. It's not like a very brain type substance, but it's something weird on the inside."

It's probably not chocolate.

Munir isn't what most people would think of as a thrill-seeker: he doesn't jump out of planes or skydive. But as an experience-seeker, he is drawn to unusual experiences and is eager to share them with others – even if they aren't along for the ride willingly. He once brought energy bars made with cricket flour to the office to share with his coworkers. And he tricked a girlfriend into eating something even more unusual.

"I think the weirdest thing that I eat is brain. Goat brain is literally my favorite dish. Growing up, it always was. I didn't really know what it was. Then one day, I opened the freezer and saw something that looked like brain – like if you imagine a brain in a cube-like flat surface. I was a little grossed out when my mom told me what it was but I got over it because it tastes just so amazing. So I actually once tricked my girlfriend – who had never tried it before and thought it was the grossest thing ever – into eating it by telling her that it was steamed cabbage. [It looks a little like steamed cabbage according to Munir.] She ate it and loved it. She said, 'This is the most delicious thing I've ever had,' and ate all of it."

"Did you tell her what it was?"
"I did."
"Are you still seeing each other?"
"No. No, we're not."

High sensation-seeking is correlated with a desire for variety in food and drinks. High sensation-seekers are drawn to unusual, exotic, spicy foods from outside of their culture. But it's more than just the experience that helps high sensation-seeking foodies try these foods. Research suggests that high sensation-seeking people have lower disgust reactions than those with average or low sensation-seeking personalities. Low sensation-seekers are more easily disgusted by foods they aren't used to. This makes it more likely that a high sensation-seeking individual will try something unusual, like goat brains or whole sautéed baby octopus or even chicken hearts.

Cy, a food blogger who writes about his experiences with unusual food, is a perfect example. "Let's see," he began, "I've had fish eyes. I've had bull testicles, but those were a delicacy in Spain. Chicken hearts was another one. I remember when I ordered chicken hearts, I got a lot of weird looks and nobody wanted to try my dish."

"That must happen a lot."

"Yeah. When they were put in front of me, they didn't look very appetizing. It was a whole pile of tiny-tiny hearts – like big peanuts or something – sitting in this stew. When I looked at the dish, I was like, 'Oh shit, this is a bad idea,' but I didn't say anything, because I didn't want to admit that this might have been a bad call in front of the other people who had already given me some flack for ordering chicken hearts. Then I took a bite. It was kind of gross at first, and then it started to grow on me. So I ended up finishing the dish."

"Did you feel disgusted at all?" I asked.

"Ah ... " there was a long pause. "No ... Once I took the first few bites, it was alright. It was actually pretty tasty." It sounded like he never even considered that he could have been disgusted by it.

Xavier Caseras and his colleagues discovered that people with a more reactive behavioral inhibition system had a greater reaction to things they found disgusting. Since high sensation-seekers have underactive behavioral inhibition systems, it might mean that they are less likely to experience disgust and emesis (nausea) reaction than their low sensation-seeking friends.[31]

Another researcher decided to test how low disgust might be linked to sensation-seeking. David Dvorak and his colleagues at the University of South Dakota conducted an experiment to find out, and the way they did it might cause emesis in you. The researchers found a video on YouTube that showed two adults vomiting on each other and then eating each other's vomit. They figured that this might produce feelings of disgust (the description alone did for me). High sensation-seekers found the videos less disgusting and experienced less emesis. Interestingly, men reported more emesis than women.[32]

Some Like it Hot

Sensation-seeking correlates with liking spicy food and even desiring the sting of a spicy meal.[33] Why do some people like the tear-inducing burn of a spicy meal? Social and cultural factors play a role. Indeed some countries are known for their spicy sauces. Additionally, early exposure to spicy foods plays a role in how much people like hot foods.[34] I'm sure that some people choke down fiery foods to prove how tough they are. Spicy foods are so popular that there is even an online calendar, the World Calendar for Hot Sauce and Chili Festivals that lists over 50 such festivals in the United States alone; from the Pflugerville Phall Chili Phest to the National Fiery Foods Show in Albuquerque, New Mexico.[35]

What makes spicy foods sizzle? Capsaicin is the active component in chili peppers and produces a sensation of burning in pretty much any tissue it comes into contact with. The more capsaicin, the hotter the pepper. The white membranes of the pepper contain the most, the seeds, hardly any capsaicin at all. Despite feeling like your mouth is being destroyed, capsaicin doesn't actually cause a burn, but rather interacts with sensory neurons. The burning sensation is measured on the Scoville Scale in Scoville Heat Units (SHU). Bell peppers contain no capsaicin and are rated a 0.

High sensation-seekers like the burn of a meal more than average and low sensation-seekers. But maybe high sensation-seekers like spicy food for the sensation itself. In 2016, Nadia Byrnes and John Hayes of the Sensory Evaluation Center and the Department of Food Science at Penn State University wanted to find out more.[36] They concocted a research plan to have people rate how much they like spicy food. Rather than buying traditional spicy foods that a person might have a preference for, since someone might like spicy foods, but not like the taste of jalapenos, for example, the researchers decided to cook up a spicy strawberry jelly using food grade capsaicin (you can find the recipe in their article if you'd like to make it yourself). There were 103 participants (only 26 men) ranging from 18 to 55 years of age.

The researchers created a soft, flowable demon strawberry jelly that came in three varieties. Plain with no capsaicin, 2 μm capsaicin (about 48,000 Scoville units), and 12 μm of capsaicin (about 192,000 Scoville units).

For reference, cayenne peppers range from 20,000 to 50,000 Scoville units and Habaneros can range from 100,000 to 350,000 units. These aren't the hottest peppers. That would be the Carolina Reaper, named by Guinness World Records as the hottest pepper on the planet. More than 200 times spicier than a jalapeno, it's rated as 1,569,300 Scoville Heat Units but can range as high as 2.2 million SHUs.[37]

The subjects were to take a spoonful of the jelly, flip it over and then place it on their tongue. After that they were to rate the intensity of the burn (no sensation to strongest imaginable sensation of any kind) and also how much they liked the jelly. What did the researchers discover? Sensation-seeking predicted how much they would like the hottest capsaicin-spiked jelly but not the moderately or non-spiked jelly. Sensation-seeking also correlated with how much they liked spicy meals, how much they enjoyed the burn of a spicy meal, and early intake of spicy meals. No other personality trait they measured was related with spicy foods at all. Their takeaway was that the burn of a spicy meal might be innately rewarding for high sensation-seekers.

But not all fearless foodies are driven by the unusual or spicy. Some try foods to be closer to other cultures. Jenny is a food critic and author of a column that features ethnic cuisine. I asked her what got her interested in writing about food.

"After I graduated from college, I went on an overly ambitious solo trip through China. I was in the western part of the country, and it's very different there than it is here. You go to these small food stalls in markets, and they have all of the ingredients laid out on a folding table – snake, squirrel, and all sorts of other weird stuff. You choose the ingredients, then they prepare it for you. There were times I ate something on the side of the street, and I had no idea what it really was. Hopefully it was something good."

"It was a little frightening for me. I am not the kind of foodie that is like, 'Give me something, the weirder the better.' I'm not the kind of person that goes and loves to eat a bunch of offal meats, but I am very open to new food experiences. For me I wanted to find the quickest inroad into the culture . . . and it's food. That's the best way you can communicate, and relate, and taste, and get a sense of other cultures . . . you get so much more information, and you're literally internalizing it, instead of going around with a guide book."

For some people, food is an experience of flavors and the opportunity to try something different and unusual. For Jenny, eating adventurously is really a kind of interpersonal connection, less about fearlessness or the specific tastes than a curiosity about the people behind the culture and food.

But there are people who are not driven purely by curiosity about cultures or exotic tastes. There are people driven by danger. Nearly all of the foodies I interviewed mentioned fugu. Served in paper-thin slices by expert chefs, fugu combines luxury with a high-stakes gamble. The intestines, ovaries, and liver of fugu (or blowfish) contain a poison called tetrodotoxin, which is 1,200 times deadlier than cyanide. The toxin is so potent that a lethal dose is smaller than the head of a pin, and a single fish has enough poison to kill 30 people. Because of the high risk, chefs must undergo two to three years of training to obtain a fugu-preparing license, and such expertise raises the price of a fugu dish to upwards of $200. But this hasn't stopped the Japanese – about 40 kinds of fugu are caught in Japan, and people consume 10,000 tons of the fish every year.

Although all of the fearless foodies mentioned fugu, Jimmy was one of the few foodies that I interviewed who had tried it. "Even as a kid I knew about pufferfish. And I'm pretty sure everyone knows that the chef has to prepare it just right or you can nick the fish and that toxin can seep into the meat and you can get very sick and even die. But if you know what you're doing, it's not such a big deal. So I read up on it before my trip and there are good licensed chefs. I was in Osaka and there's numerous restaurants that specifically focus on fugu. I got a sample platter and it was served five different ways – in a soup, raw, and cooked. It made my lips tingle a little bit, but it seemed relatively safe."

> "What's it taste like?"
> "Like flounder." He sighed. "Not that exciting. I was actually a little let down."

If you are not a foodie, or a high sensation-seeker, you may ask yourself why someone would order strange food, that borders on being disgusting or one that might actually kill you? If you've ever wondered why someone would routinely order the strangest or even most disgusting thing on the menu, you might be asking the wrong question. The better question might be "Why not?"

Remember Cy, who ordered the chicken hearts? When I asked him what his motivation was he said, "Most of the time when I order something that's a little off the wall, the rationale is like, 'Screw it. Why not?' I mean, what's the very worst that could happen? I'm in a restaurant. There are people everywhere. What could possibly go wrong? I'm in a restaurant, so what could happen? There's not much else to it."

"I also think there's an allure to trying different ingredients I've never tried before. That's pretty much the only reason I do it. Like when I ordered pig blood stew. I remember the first bite I had was kind of gross. There was a lot of – it had a metallic, iron-like taste to it. I just told myself that I had to take a couple of more bites, and then after that initial shock, it kind of tasted pretty nice. I've definitely had pig blood or pork blood in other dishes since then. I must not have hated it all that much."

> "Did you finish it?"
> "Yeah, I finished it."
> "Is there anything that you wouldn't eat – within reason, of course?"
> "No, not really. Once I ate fish eyeballs. Anything goes."

Once you've tasted fish eyeballs, there's no going back, I guess.

~

It's clear that high sensation-seeking can impact nearly every aspect of a person's everyday life, from food to jokes, art, media, music, multitasking, and travel.

I checked in with the White Rabbit about a year after I first contacted her to see how her journey went. She did it. 300 days without spending any money and she visited 14 countries all around the world. No doubt her high sensation-seeking personality not only inspired her trip but also kept her safe from some of the dangers to which she was drawn.

You never know where the white rabbit might take you, just ask Alice.

For some high sensation-seekers the white rabbit beckons them beyond wanderlust. It tempts them off mountains, around bends, off cliffs and even into the mud.

4 LIGHTS, CAMERA, ACTION: SPORTS AND ADVENTURE IN HIGH SENSATION-SEEKING

Matt Davis was a self-described everyday guy. "I was your average 40-year-old, married with a couple of kids kinda guy that played neighborhood softball and an occasional flag football game. No one would call me obese, but I certainly wasn't in shape. I couldn't run more than a mile." He did his first obstacle course race, the Warrior Dash, because it sounded fun, it seemed manly, and you got a goofy hat and a turkey leg. He wasn't looking to change his life; his sensation-seeking personality drew him into activities with a high adventure factor.

An obstacle course race (OCR), also known as a mud run, is an adventure endurance race. Swamp Dash and Bash, Tough Mudder and Beast Race are some of the more popular ones. Those who pay to participate can look forward to the muddy mayhem, arctic enema, cry baby, and more. These are the affectionate names mudders give to such activities as belly-crawling underneath barbed wire with your face in the mud, sliding feet first into a pond of icy water, and crawling through a tent filled with tear gas. It all sounds like a scheme that might have been concocted by Dr. Evil to thwart Austin Powers. I guess the sharks with lasers attached to their heads were already booked.

OCRs attract a variety of sensation-seekers, including Matt. Matt told me about his first experience with these races which was back in February of 2012. "They had us do what's called 'walk the plank,' which is jumping off a fourteen-foot platform into a lake. I remember being scared out of my mind. This military guy was up there, and he said, 'I'm going to count to three. When I get to three I'm going to push you, so you should probably jump.' I don't know if he was serious or not, but I jumped, hit the water, came up, and I didn't die. I let out a yell, because I felt alive like I had never felt before. I thought, 'Now I can do anything.' I ran twelve miles. I survived a freezing day. Then I jumped off a fourteen-foot platform into a lake. I felt invincible!"

After the Warrior Dash came the Tough Mudder Extreme Trail, then the Spartan Death Race; he was hooked. "I think it happens to a lot of people. I felt a sense of accomplishment I hadn't felt before. I felt re-engaged with fitness, re-engaged with my youth, and I was looking to sign up for the next one." His life started to change and he felt inspired by the people doing the races and the employees and volunteers setting them up. He started a podcast, a website, and wrote a book: *Down and Dirty: The Essential Training Guide for Obstacle Races and Mud Runs*.[1] He's now the cofounder of Obstacle Course Media, a company that produces and promotes obstacle course races, and his weekly podcast is the most popular OCR podcast out there. Matt lives and breathes OCRs.

Many OCRs like Tough Mudder aren't actually races in the traditional sense. No one is timed and there's no winner. Those who complete the races are awarded medals, turkey legs, beer, bananas, and bragging rights. The lack of competition encourages camaraderie; helping others complete the course is a huge part of the experience. The source of pride is crossing the finish line.

It's difficult to estimate when OCRs started. Some might say that gladiators were the original tough mudders. Modern obstacle course races probably started back in 1987 with the Tough Guy Race near Wolverhampton, England, according to Matt.[2] Since then, it's mushroomed into a multi-million dollar industry with over 4.5 million participants worldwide in 2015.[3] There's little wonder why. People who complete OCRs report being fundamentally changed by the experience. As Matt put it, "I think it changed me so that my DNA is different. I act differently now. I've jumped off that fourteen-foot cliff a ton of times. Of course, I'm always scared, but it's not as scary as it was the first time."

I knew what he meant, kind of. One of the reasons high sensation-seekers are attracted to events like Tough Mudder is due to their *arousal potential* – a concept that describes the ability of something to draw your attention. It's one of the reasons high sensation-seekers get used to horror movies so quickly.[4]

But people involved in these types of races do get injured, sometimes permanently. After all, many people enter OCRs without adequate preparation for such unusual physical challenges. At one event alone, over two dozen people arrived in a local emergency room with burn marks from tasers, inflammation of heart muscle, and fainting. Most were treated and released; some were admitted to the hospital, and at least one went into intensive care. A few required ongoing treatment for months.[5] But the majority of runners avoid injury and finish wet, muddy, and with an enormous sense of personal accomplishment, which may be one of the reasons high sensation-seekers are typically involved in more leisure activities than the rest of us.

Researchers polled sensation-seekers and asked them to list the things they do for recreation. Not only did they report a larger variety of activities than most, but also their choice of activities was flavored by their quest for thrill and experience. When it comes to leisure activities, high sensation-seekers try more, travel more, and have less anxiety when they encounter risky or unusual situations.[6]

People's leisure activities, what they do when they can do whatever they want, tells you a lot about them. What I noticed about high sensation-seekers over and over again is that they tend to spend their leisure time in ways that don't seem all that leisurely. Crawling under barbed wire in the mud or plummeting dozens of feet into arctic-cold water does not seem like a relaxing afternoon. Tough mudders are fully aware of this, but they do it anyway. The question is, why?

It's been widely established by researchers that high sensation-seekers love high risk activities like skydiving,[7] auto racing, and hang gliding.[8] Christopher Cronin and his colleagues administered Zuckerman's Sensation-seeking scale to two groups: members of a university mountain climbing club and a group of psychology students not involved in mountain climbing. The members of the mountain climbing club had higher total sensation-seeking scores. They also scored higher in the experience-seeking and thrill- and adventure-seeking subscales of the sensation-seeking scale.[9]

The higher your total sensation-seeking score, the more likely you are to try higher-risk activities.[10] Low sensation-seekers stick to lower-risk activities. High sensation-seekers like low-risk activities too – BASE diving as well as ballroom dancing. Risk isn't the goal; it's what they must do to have the high sensation-seeking experience. Those with high sensation-seeking scores also enjoy all sorts of non-risky but stimulating activities.

High sensation-seekers also biologically experience their edgy hobbies differently than low sensation-seekers do. High sensation-seekers' lowered cortisol and heightened dopamine, and sometimes lack of fear, not only draw them to certain sports and pastimes, but also can make them experts in their execution. While low sensation-seekers might be flooded with sensations during hang gliding, a high sensation-seeker's neuropsychological makeup makes it easier to perform the tasks that high-risk activities demand.

High sensation-seekers' edgy hobbies can give them a sense of mastery over themselves and their environments and can make them feel really alive. Their hobbies can also be dangerous and costly when they're taken too far.

Roller Girl

High sensation-seekers perceive risk differently than low sensation-seekers – so much so that they actually enjoy the riskiness of their hobbies. They like relying on their instincts in dangerous situations and they enjoy the focus and increased self-control they get from their activities.

Pam, a young professional currently living in Texas, told me about the high-stakes, street-style roller blading she likes. She recalls with fondness her identity as "Roller Girl," which sounded a bit like a superhero alter ego, and a particular memory from when she was living in Chicago. "One of my memories is that we were going down by Michigan Avenue around the art museum, down in that area. We would go flying in front of a taxi and then flying in front of a bike. We'd go flying up on the curbs and then jump down and then go in the street. A bike would pull in front of us and right away I would see it. It's almost like all of sudden he sees me but I had seen him 20 feet back. I could almost tell that he was probably going to hit me by just the angle of the bike, the way he was riding, everything. It was just an unbelievable moment of focus and concentration.

"We're flying down this hill and I'm like, 'Sheila, I don't know how to brake.' She's like, 'Well, you better brake or you're going to die.' I wasn't afraid. That is the most interesting thing. All of a sudden, I was like, 'Okay, I need a target.' I found a light post. I was like, 'I'm going to aim for that.' We're flying down the hill. People got out of our way and I grabbed the pole and rolled around it three or four times. I'm laughing. Sheila is laughing. Everyone is laughing. And I didn't die. There was this feeling of zero fear ... you have no fear. You actually get more calm and more focused and more centered. It's the opposite. It's really weird to describe it. It's a sensation that when you're just in the movement and in the moment, there is no place better on Earth. No place."

Pam's ability to navigate the streets of Chicago was heightened by her ability to not become overwhelmed by the sensations she experienced. What's more, as a high sensation-seeker she probably perceived the activity to be less risky than it actually was, while average and low sensation-seekers may see these activities as way too risky. We know that high sensation-seekers interpret the world a bit differently and they see the environment around them as less risky.[11] They simply don't view the activity as dangerous, risky, or threatening in the ways in which their lower sensation-seeking friends would. Remember the approach and avoidance systems we discussed in Chapter 2? Because their bodies produce fewer stress hormones during high-risk activities, and the "stress" button does not get pushed, they may think situations are less risky. They don't feel stress in the same way.

You might think that high sensation-seekers' penchant for risky ventures makes them more likely to have injuries, but one study of high school athletes found that sensation-seeking was unrelated to injuries.[12] Another group of researchers studied patients undergoing physical rehabilitation. They didn't find more high sensation-seekers in that group, either.[13] The better you are at something, the less dangerous it's likely to be. Risk is related to skill. Perhaps people with high sensation-seeking personalities have fewer accidents because they have a higher skill level in the risky activity. What's more, high sensation-seekers tend to bounce back more quickly from injuries and are more likely to resume the activity in fewer days (perhaps the pain doesn't bother them as much).[14] On the other hand, low sensation-seekers tend to be more stressed by injuries and more likely to abandon the activity after an injury.[15] So not only do high sensation-seekers perceive risk differently, but also their experience of risky activities tends to be less, well, risky.

Safety First?

You might think it would be better for Pam and other high sensation-seekers to don safety equipment like a helmet. It would certainly make Pam's skating adventures safer. Or would it? The helmet may actually not help as much as you would think. It seems that safety devices may actually increase a person's possibility of risk-taking.[16] Making the activity safer doesn't always make the activity safe. Your perception of safety can influence how much risk you take. This is known as risk compensation.[17] It means people take more risks when they have equipment that protects them. It's been demonstrated with drivers using a car with or without safety devices,[18] with kids in an obstacle course with or without safety gear,[19] and even with cyclists going down a steep hill with or without helmets.[20] In all of these cases, people with more safety equipment took greater risks than those without. This might explain why people with the latest self-driving robot cars fall asleep or even climb over to the passenger side to take selfies.[21] Sometimes safety makes people complacent.

The Crux of the Problem: The Real Reason High Sensation-Seekers Seek Adventure

Timmy O'Neill has an extensive resume of thrill-seeking adventures – from backcountry skiing to kayaking. Timmy has made YouTube videos of himself scaling buildings in L.A. He's speed-climbed El Capitan, kayaked the length of the Grand Canyon seven times, climbed buildings without ropes, and he's taken a few hits for it. He survived a 120-foot fall in 2000 while climbing a granite formation in Pakistan. Timmy is also a professional comedian, as well as a philosopher of thrill. He told me that the reason he attempts these high-risk activities was what he called "the crux."

The word *crux* has several definitions. A crux is a puzzling or difficult situation. It's also a crossroads. However, it also means "an essential point requiring resolution," and according to O'Neill, it explains why people engage in risky thrill-seeking experiences. Thrill-seekers put themselves in dangerous, shocking, and wondrous situations, specifically because it forces them to find their way out, down, or through.

This fear can be both a lock and a key. "Fear can prevent us from doing the things we love and loving the things we do, but it can also, if we are lucky, help us to access peak emotional experiences," writes Florence Williams in *Outside* magazine.[22] Many high sensation-seekers are extremely lucky in this way. "I never felt as alive as when I was so close to not being alive," one high sensation-seeker told me.

One adventure sports professional told me about a situation in which he needed to react quickly and instinctively. That situation, he said, made him feel excited to be alive. "I was on the coast of California, there's this hike, it's called the Lost Coast. The mountains butt up against the coastline. The only way to experience this part of the coastline is to hike 24 miles by foot. You can't drive up to it because the mountains are so jagged. What happens is that the tide comes in and out throughout the day, twice a day. As you're walking along the coast if you don't position yourself correctly the tide will come in and totally surround you. You have to get to high ground very quickly."

I tempted fate, I pushed myself a little too hard and I was in a situation where I had to climb an 80-foot eroding cliff, where you're grabbing onto roots and things like that. I had to take off my backpack and throw it up a couple feet above me and then try to climb up. Meanwhile the ground is literally falling out from under my feet. If I were to fall in that situation, or get hurt, there's no help. There is a moment where you're going 'okay, hold on.' What I've learned is that you can't let it paralyze you because that's when you get hurt. You have to almost have fun with it. There's this personality in me that will just turn on, and I have a smile on my face. All right, okay, let's do it. You have to be excited to be alive, it turns that on to be in those situations. I would never want to have to experience that again, but when I was in that moment I wasn't going to shy away from it. I was just going to be in that moment and take it how it was going to come. Then you survive and you look at it and go okay cool, I can move on. Let's not do that again."

Life-or-death moments like this one came up in many high sensation-seekers' stories. Whether it's street luge racing, sky surfing, or parkour, there's a right way to perform these risky activities, and deviation from that way can result in dire consequences. They fight the elements, the mountain, distractions, and sometimes their own fear in order to be at the crux. They crave being in puzzling situations – essential points that require resolution. "For example," O'Neill said, "Say you're on the side of a sheer granite wall, and you have to use your body in a very precise manner to be able to overcome a difficult expanse. If you don't execute it in just the correct way, you'll fall. Gravity is a tough teacher. But there is a reward. You're forced to problem solve, remain calm, and be self-reliant and resilient – all great qualities that people value."

The crux is heightened even more by the possibility of the unexpected. We can take Timmy's example a step further to see why. Rock climbing has dozens of unexpected risks: falling rocks, falling from rocks, falling into rocks, losing your footing, losing your grip, you really don't know what's coming next. An experienced rock climber must know just what to do in totally unexpected situations and execute it quickly without thinking about it too much. He or she must solve the puzzle rapidly, without much hesitation. "Analysis is paralysis" is a common expression among adventure sports participants. You have to react instinctively in these high-risk situations or face the consequences.

For many high sensation-seekers, the buzz comes from placing themselves in these crux experiences and emerging whole. If you think about it, even low-sensation-seekers do this all the time. From crossword puzzles to knitting, to kiteboarding, we seek to challenge ourselves and find reward in the challenge. It's not a worthwhile challenge if you know you can do it. For some high sensation-seekers, the more complex the challenge – the higher the risk – the higher the reward.

Plus, they say, it's fun.

Not everyone has the skill or nerve to attempt the highest risk activities. For them, moderately risky activities can serve the same goals. Body-contact sports such as soccer, free climbing, and karate are common medium-risk hobbies. These hobbies don't necessarily involve danger, but a few bumps and bruises along the way are part of the experience. Those with high sensation-seeking scores are drawn to these activities more often than many low sensation-seekers.[23]

Beyond Human Limits

Katrina Pisani can't remember a time before Science North. "I grew up with Science North in my life all the time, and I loved coming here as a kid," she explained. "I used to come with school, with my parents, and with my friends and used to do all kinds of things here. I always thought it'd be so cool to work here. It's such a neat place."

Science North is Northern Ontario's most popular tourist spot, bringing in over 300,000 visitors in any given year.[24] Not bad for a city of only 160,000. It consists of two snowflake-shaped buildings just south of the central business district. Inside is a science funhouse that will delight pretty much anyone. Science North is also involved in conceiving, designing, and building traveling science exhibits. Currently they've created a half-dozen or so traveling exhibits that have been featured in over 100 locations around the globe. If you've visited a science center or science museum there's a good chance you've seen one of their exhibits.

Katrina has a background in psychology and came to Science North after graduate school as an intern and never left. Her main job? She's involved in evaluating the experience of visitors. When research indicated that visitors would like to see an exhibit focused on extreme sports, Katrina was asked to work on the concept plan because of her background in psychology. She also happens to love extreme sports, so she jumped right in. Soccer and flag football are her core sports and she works out most every day: kettlebell, weight lifting, boot camp. "Anything that's super challenging," she said with a smile. "People always laugh at me but I always say when I work out that at the end of my workout, I should feel like I want to throw up."

As they built the exhibit, they soon realized that extreme sports didn't really capture what they wanted to describe. "We got together in groups and would pick our favorite words to describe the exhibit. They were things like 'practice' and 'training' and 'perseverance' and 'motivation' and 'highly skilled.' It was interesting to see that it moved away from the idea of extreme sports and moved into the idea that these individuals are in it for something more than just the extreme part of it. They're in it for how it makes them feel and the intense training that it takes and the motivation. It just creates a different lifestyle and a different way to think about risks and a different way to perceive sports."

That progression of thinking is not unusual. When many people think about extreme athletes, they think about people who are, in some ways, thoughtless. But the more you dig into it the more you discover that for the clear majority it's the opposite. These are highly skilled professionals.

"These are amazing, thoughtful, purposeful people who are not about the adrenaline rush," said Amy Wilson, who is a video editor, video producer, multimedia producer, editor, and sound mixer at Science North.

Amy got involved in the project when it was decided that one of the first things they wanted for the exhibit were interviews with athletes. Unlike most of the Science North team she doesn't have a science background. "My last science course was way back in high school," she recalls. Amy is a storyteller and helps weave together the science stories.

Amy's first task? Hit the road and interview the thrill-seekers slotted to be featured in the exhibit, and like many people who dive into this concept, her idea of sensation-seekers changed over time. They weren't necessarily the adrenaline junkies she expected them to be.

Amy trekked across the United States and Canada interviewing the athletes and probing into their motivations for what they did and why. She also asked about how their personalities may have helped or hindered the process. Three athletes she met who exhibited qualities or traits that are common in high sensation-seekers are Will, Katherine, and Jeb.[25] Their high sensation-seeking personalities allow them to focus, persevere, and bust through the crux experiences they face in the extreme sports that are part of their lives.

Will Gadd: Focus in the Frost

It's easy to think of extreme sports athletes as being impulsive and reckless. Maybe because they often attempt things that many people would only do with a leap of faith. But it's important to remember how much focus, attention, and training it takes to accomplish them successfully. Take Will Gadd. Will is a paraglider pilot who once held the world record for distance in a paraglider: 263 miles in Zapata, Texas.[26] He's also an ice climber and in 2015 was the first person to climb up Niagara Falls.

Niagara Falls is arguably one of the most famous waterfalls in the world. The falls, which straddle the border of Canada and the United States, welcome some 20 million visitors a year. Some 3,160 tons of water flow over Niagara Falls every second.[27] Dozens of professional daredevils have traveled down the falls. Will was the first to travel up. Why Niagara Falls? "Because it's the coolest waterfall in the world that ever has any hope of freezing," Will explained. But it's no easy task. "I have to figure out how to operate in an environment that's pretty much out to kill me."

After working with the New York State Parks Department and the New York State Parks Police, Will and his team set out to create a plan for a safe climb that wouldn't damage the falls themselves.

Importance of Focus

Lots of things make the climb treacherous. "Niagara Falls is massive. You've got semi-trucks of water coming over the edge every second. There are trucks and trains and office buildings pouring off of that thing. And if you're in the way, you're going to get killed." The first part of the journey is to travel over what Will affectionately called "The Cauldron of Doom" (it's where the falls gush into a massive hole in the ice). From there Will picked and climbed up the frozen falls stopping every few feet to add ice screws and other pieces of climbing equipment. And that's not all. The air is filled with mist, which freezes to you. It's just an astoundingly complicated place to be. And then there's the ice.

"It's not easy ice, either," Will added. The ice in Niagara Falls forms in layers. There's a layer of ice, then fluffy air-filled snow, then more ice. Not at all like the smooth ice of a hockey rink. Normal ice flows down like ice on your sidewalk. But Niagara's spray forms ice in layers that are inherently unstable. The night before the big climb Will dropped in from above to clear the route of the most dangerous hanging ice that could break off (with him) during the climb. "I was taking off pieces the size of small cars."

One of his main climbing tasks is to try to understand where the good ice is and where the bad ice is. Try to imagine hanging off the side of an ice-covered waterfall with ice climbing picks. Some of the ice is solid and thick and will hold your weight as you pull yourself up. Other ice will slide off once your pick goes in. That's the ice you want to avoid. One false move could be deadly. Will has to focus, listen, and choose wisely and decisively, and his high sensation-seeking personality is ideal for this task. Lower amounts of cortisol and increased flow allows Will to pick the perfect ice. And there's a lot to figure out.

"There's two things you use to figure out how good ice is. First it's the sound when you hit it. What's it sound like? What's going on? Does it sound like you just punched a thin shell over something soft behind it? If you hear, 'Crk', that's no good. Or does it sound hard, and clean, and 'thunk thunk'? And that's a really good sound. My mind is always listening very carefully when I'm climbing, to what's going on."

The second is how the ice looks. "Is it aerated?" Meaning does it have a lot of oxygen in the ice, and will it be less stable, or is it hard and blue? Unfortunately, most of the ice at Niagara is not hard and blue. It's filled with air that makes it soft and easy to break. "I'm trying to find ice that is somehow well-bonded to the rock underneath it. And I'm looking and listening for the sounds trying to find ice that isn't going to fall off."

It's nearly impossible to wrap my head around what the experience would be like. Even when they're not frozen, the falls are monstrous. It seems like climbing them would be terrifying. Yet fear wasn't on Will's radar. This kind of focus and execution isn't unusual for high sensation-seekers. Their ability to be unruffled and focused allows them to act and rely on their training.

By the time he started climbing Niagara Falls, Will had done a tremendous amount of work to be there, and he said he wasn't afraid. "Fear is usually a sign to me that you're in a place," he explained, "maybe that you don't understand, or that you have legitimate worries. If you're afraid, then something needs to be addressed or changed. And by the time I started climbing Niagara Falls, I had real concern, and probably still some pretty good adrenaline. But the fear was gone. Now it's time to act and to execute and to do something at a high level and do it right."

But real life is always different than the plan.

"The ice was much worse than I expected. The falls were much steeper than I thought they were going to be. I thought it would be more vertical or maybe slightly less than vertical."

Will picked his way up 148 feet (about the height of a 15-story building), and once at the top he let out a victorious yell.

"People often think that ice climbing is mainly a physical sport. That's the easy part," Will explained. "It's all the mental stuff. And staying on top of that, it's much more difficult. And then finally having fun, I wouldn't do any of this if it wasn't fun, right? This is pretty awesome."

A Lousy Drug

Many people might consider Will to be an adrenaline junkie who is in it all for the thrills. He sees it another way. It's part of who he is and honestly adrenaline can get in the way.

"I grew up in a family that hiked, climbed and went into the mountains whenever they could. Some of my earliest memories are of long backpacking trips, wind-blown summits and surviving winter skip trips. The first sport I really got into on my own was caving. When I was 14, I started kayaking. At 16, I bought my first climbing rope, and did my first new rock route. At 25, I first flew a paraglider. At 41, my daughter came into the world. She's already doing all the stuff I did as a kid, and she's an athlete too."[28]

But Will doesn't see adrenaline as his friend.

Adrenaline, sometimes known as epinephrine, plays a major role in the fight, freeze, or flee response. Adrenaline has a broad influence over many organs in the body. For example, adrenaline will increase blood pressure, heart rate, pupil dilation, and blood sugar level. It's a great tool to help in situations where you need a burst of energy. It's no wonder many people think that high sensation-seekers are adrenaline junkies. However, too much adrenaline can be disruptive when focus is key.

"Adrenaline's all about dosage," he says, "so if you're getting a huge hit of adrenaline, it's actually a really lousy drug. If you've ever almost had a car accident or almost fallen off something high, then you know what adrenaline feels like. You feel nauseous, it's this spike of stuff that's designed to get you out of a bad situation. That's kind of the role of adrenaline. At a very low dose it's interesting. It switches your brain on. It fires you up physically, cuts out distractions. It's a pretty useful drug at low doses, but at high doses it's lousy, so you have to figure out where on that continuum your brain works well and then get into environments that do work well. I've tried BASE jumping. It's too much pure adrenaline. I don't like it.

"But I think most people think people in my shoes, we're looking for that huge hit of adrenaline. We're not. A huge hit of adrenaline is a lousy drug. What I'm looking for is to do really difficult things and do them well. That's what's challenging to me. That's what fires me up in life, whether it's putting together a big project and figuring out how to make it work or climbing well. It's solving puzzles and doing really interesting, cool things that are exciting and fun. That's what life's about.

"If I just wanted adrenaline, it would be a lot simpler. I would just go run back and forth across the highway. If that's what's making my mind tick, there are way, simpler ways to do it."

Tuomas Immonen and his team suggest that some people, including researchers, are too "risk-centric" when they are describing the kind of sports Will does. Most people, they suggest, describe the kinds of experiences that extreme sports produce as "positive, deeply meaningful, and life-enhancing" and these have little to do with risk or adrenaline.[29]

Katherine Beatlie: Perseverance and Grit

It seems to me that every few years they come up with a brand new way to get hit by a car. In the 2000s it was the Segway and Razor Scooter, and in the 1980s it was rollerblades. I remember the very first time I put on rollerblades. I certainly never expected to match the skill of Roller Girl who you read about earlier, but I thought maybe it would be fun to skate around the park. How hard could it be? Turns out I somehow injured myself while simply putting on the skates and decided to abandon the activity forever even before standing up. High sensation-seekers aren't like that. We know from the research we discussed earlier that high sensation-seekers are more likely to press on even when they do get injured.

This is certainly the case for Katherine Beatlie. By day Katherine is a television scriptwriter in Los Angeles but she's always been a high sensation-seeker.

"When I was a kid I was always trying to ride bikes and skateboard and I always wanted to surf and bodyboard. If it had wheels I wanted to do it. And I was terrible at it." She laughed. "I could hardly make it move and I was so bad but I loved it and I kept at it." Yes, unlike me, Katherine kept at it. Psychologist Angela Duckworth calls this grit.[30]

As a teacher, Duckworth was puzzled about how some of her students performed on their assignments and exams. Some of the strongest students didn't necessarily have high achievement scores, and some of the brightest kids weren't doing all that well. It seemed that there was no connection between how "smart" the students were and how well they were doing in her class.

Angela left the classroom and went to graduate school in psychology. Her big question? Who is successful and why. She looked at elite military training graduates, rookie teachers, even national spelling bee champions to see if there was something common about who succeeds. One thing stood out as a predictor of success – grit.

Angela Duckworth says that "Grit is passion and perseverance for a very long-term goal. Grit is having stamina. Grit is sticking with your future, day in day out, not just for the week, not just for the month but for years and working really hard to make that future a reality."[31]

The truth is that lots of very talented people don't end up doing the things they need to be successful and that talent alone isn't related to grit. She suggests that talent counts once and effort counts twice for success. According to Duckworth "Grit is about having what some researchers call an 'ultimate concern,' a goal you care about so much that it organizes and gives meaning to almost everything you do. And grit is holding steadfast to that goal. Even when you fall down. Even when you screw up. Even when progress toward that goal is halting or slow."[32] If you want to know how you score on Duckworth's grit scale, take a look on her website or go to Appendix 4.

Katherine's ultimate concern is her quest to flip and glide. "I used to kneeboard," she explained. "I also had a big skateboard that I would ride on my knees." But since Katherine had mobility concerns she needed special equipment to get to the next level. Then she met Mike Box who manufactures wheelchairs, but not just any wheelchairs. The chairs need to withstand the scuffs, spills, and stunts of their riders. Wheelchair motocross, also known as WCMX, requires a rugged chair and even more so a rugged constitution. Katherine got her WCMX chair in 2012 and the very next week she was at a skate park in Santa Monica dropping into bowls and never looked back.

Her favorite trick to do in WCMX is the back flip. In fact, she's the first woman to ever do a backflip in WCMX. Katherine described the first time she ever did this. "All of a sudden you're in the air just pulling your chair back and getting it back underneath you. To me there's nothing like that feeling of seeing the ceiling and all of a sudden seeing the floor again, but this time it's underneath you." And she loves the challenge despite how much effort it took.

"I got this idea that I wanted to do a back flip, but I really wanted to be the first girl to do it, so just being able to set my mind to that and accomplishing the backflipping, being the first one to pull that off, was a huge thing and that's always something that stands out for me. Yeah. It took me years just to get up to the point where I was ready to try it, and the thing about a back flip is you could just go out and try a back flip if you wanted, if you're crazy, but really what you need is a place something like Woodward West here, where you have a foam pit where you can practice." And practice she did.

"It was definitely many days of trying to figure out that physics of when to pull and how fast to go and where to look. It's one of those tricks that it's probably the one that I put the most work into, which is why it's the most special to me.

"The first time I got my wheels under me, I was not expecting it. I go to do these tricks and I fully expect to fall, every single time, so when I pull it off, I have a moment of 'Wait a second. Um, OK. Wait, I'm sitting upright, I'm not on the ground!' In that eight months, I decided that I'm not going to give up on this, I'm going to stick with it even though it's slower than I thought it would be, and even though it hurts, and I'm tired, somebody else might get it before me. Just keep going, 'cause it's what I really wanted.

"If it's easy, I don't see the fun in that and there is less of a reward. It feels so much better when you do something that's hard and finally land it. You might be able to do something very gracefully, but if it's easy, what's the fun in that? So I'm always looking for that next thing that's going to challenge me."

While many high sensation-seekers certainly have grit, not everyone feels grit is the secret weapon to success. In 2016, Kaili Rimfeld and her colleagues discovered that grit wasn't a big factor at all for academic achievement in the 2,321 pairs of twins they studied.[33] Other studies have found similarly disappointing results.[34] While the concept of grit alone may not explain Katherine's success, her passion and sensation-seeking certainly make a difference.

Jeb Corliss: Rising above the Crux

When you think of a thrill-seeking extreme sport it's easy to think of wingsuiting and people sailing over and through the landscape. Wingsuits work by adding surface area to your legs and under your arms to increase lift. They give you wings. You end up looking like a giant flying squirrel. But you don't just leap into the air from the ground. Wingsuit flyers jump from airplanes or BASE jump from cliffs or mountaintops. After sailing through the air they deploy a parachute to land safely.

Jeb Corliss is a professional skydiver, BASE jumper, and wingsuit flyer. He has jumped from sites including the Eiffel

Tower in Paris, the Space Needle in Seattle, the Christ the Redeemer statue in Rio de Janeiro, and the Petronas Twin Towers in Kuala Lumpur, Malaysia. If you Google "Jeb and wingsuit," about a dozen or so videos will pop up in your search engine including those of him flying through Tianmen Cave in China and Ball's Pyramid in Australia. All are death-defying stunts with wingsuits.

Jeb became famous for doing massive mega stunts with the wingsuit, especially in China. According to Jeb, when he first went to China no one knew who he was and wingsuits were such an oddity that there wasn't even a word for them. "They actually had to invent a word for us to explain to the population what we were doing." The producers didn't have high hopes that people would watch it or care. They thought if 10 million people watched, it would be great.

The estimates were a little off. Well over 500 million people watched Jeb's first stunt live. So many that they had to stop airing it in parts of the country because people stayed home from work to see it.

Jeb says he loves extreme sports because they helped him "find himself." He started out BASE diving and at the time he felt lost. "I was one of those 16-year-old kids who didn't know who I was or where my place was in the world . . . I didn't have a purpose, and I was suffering from a heavy depression and BASE jumping was something for me to wrap my life around."

For Jeb the training involved in BASE jumping gave him focus and he learned a lot about himself. "I learned that the sport wasn't what I thought it was. The sport is a psychological test. It's you finding out who you are, it's finding out what you're made of, what you're capable of, how far you can go before you break. Now, at 41, I know exactly who I am, and I know exactly what I'm capable of, and I'd have to say my favorite aspect of the sport was that, the discovery, the learning about who I was.

"I love activities that force you to get up off the couch and out in the world to see new things and experience life. For me, that's very important. That's why I like surfing. That's why I like climbing. That's why I like skydiving. That's why I like BASE jumping. That's why I like scuba diving. All of these activities have forced me to travel around the world, meet new and unique, interesting people and cultures."

What doesn't he like about it? Adrenaline. Like many professional extreme sports athletes he abhors adrenaline.

"I would have to say when I was younger, like probably my early 20s, I did enjoy the feeling, like the rush you would get from doing these kinds of things, but as time went by, I started getting kind of a negative association with adrenaline. I don't like the way it feels. I don't like being scared. I don't like terrifying myself anymore. I do not like the feeling of adrenaline. I think adrenaline is actually a nauseating sensation, and it actually makes me feel sick to my stomach. You get that feeling of your heart coming into your throat and your whole body becomes super hardwired, and you become just so hyper-focused. Honestly, I don't like it. I don't like it at all. If I could do what I do without it, I would, but unfortunately, that's just not the way physiology works."

He compares his motivation to the motivation of astronauts. "They didn't go to the moon to see how much adrenaline they could get. They had an incredibly difficult task, and their goal was to get to the moon and back in the safest way they possibly could. That's my goal now. I do select incredibly dangerous things to do, but my goal is to try to get in there and get back in the safest way I possibly can. Sometimes I succeed. Sometimes I don't."

Table Mountain

In 2012, Jeb was flying wingsuits off Table Mountain in South Africa shooting for a show called *Real Sports* on HBO. He did a week of jumping for the show and when it was time to wrap up, he had a little extra time and decided that he would do a few extra jumps, just for fun. Since he was training for another project he had some friends set up balloons to use as targets that he could fly toward. "I was talking to my buddies who set up the targets, and they told me, 'Jeb, it's kind of windy. You know, be careful. We don't suggest you go for the balloons.'"

"Well, I'll tell you what. If I see the balloons moving, I won't go for them, I'll just fly over them," he responded.

Looking back Jeb realized that ignoring the balloons was harder than he thought at the time.

"When you're moving 120 miles an hour and you're looking at an object that's as big as a balloon is not really easy to do.

"I'm just so focused on the balloon that I'm thinking, 'I'm getting that balloon.' And I did. I hit the balloon. The problem was I hit the cliff before I hit the balloon. I hit the ledge, clip the balloon, and then start tumbling off the edge of the cliff. As I start tumbling off the edge of the cliff, the first thought that went through my mind was shock. I was shocked because I didn't see it coming, I thought I was good.

"I realized instantly what had happened. 'How am I not dead?' My thought process has always been if you ever impact at terminal velocity, you die instantly. Many of my friends had died that way, I watched people die like that."

He couldn't wrap his head around the idea that he was still alive. "Then my brain did this strange separation. One part of my brain was pure math. Pure science. Just technical: 'You're tumbling, recover.' I'm judging altitude and angles and everything I need to do to open a parachute and survive."

The second part of his brain was much more philosophical. He thought to himself, "Well you're dead. There is zero chance of survival. You have just impacted into flat, solid, granite at 120 miles an hour. You are going to die. So, how do you want to die? You can pull your parachute and maybe get 30 more seconds, maybe a minute, maybe ten minutes. And then heli-rescues take around 2 to 3 hours." He knew what was ahead. "So, you have a choice." He tells himself, "You can pull and have a slow, agonizing, painful, just suffering death. Or don't pull and just let it be over. I was struggling, I was struggling with it. I mean if you know death is coming, are you going to take the suffering?"

But the mathematical part of his brain and his high sensation-seeking personality was calculating and in survival mode. "Pull now or you die. This is it. This is the moment. Right now. I remember consciously thinking, 'Ah, good thing you like pain. Let's see how much time I can get. That's it. Good thing, you like pain.'"

And he pulled his parachute.

He hit a second cliff and then things got fuzzy. "I remember being in extreme pain. Extreme. And then it was 120 degrees, record-breaking temperature. I'm all in black, I was getting cooked alive."

"I've died a few times in the past so I know what it feels like when your body starts just shutting down on you and I was dying for sure. And all of a sudden, the helicopter comes. I remember being pulled into the emergency room and I remember hearing them say this is going to be a double amputation. They were talking about my legs. And I think, 'Oh, that's great.' Then I realized what that meant. If they cut my legs off, that means there's a chance I won't die."

Soon after there are a series of x-rays and other procedures. Then finally the doctor walks into the room to talk to Jeb.

"You are the luckiest man on the face of this planet," the doctor says.

"Really, why?" Jeb asks.

"Well, you have really strong bones. It's unbelievable. You have a broken fibula. You have multiple broken toes. Both your ankles are broken. You have a massive gash in your shin that's going to require skin grafting and multiple surgeries. You've degloved both your thighs. The muscles have been torn from the bone and been put into your knees which is lovely. But aside from that, you're going to be okay. You're going to have a 100 percent recovery. You'll be able to walk again. You'll be able to go back to your life."

"And it was such a strange feeling, to know you're going to die, and then to think you're going to lose your legs, and then to be told you're going to be just fine, you're going to have a total 100 percent recovery. From a completely unsurvivable accident, something you can't survive."

Bouncing Back: The Flying Dagger

Fast forward to a year later. A year of recovery and Jeb is back in China scouting out a location for his next big stunt. He called it "The Flying Dagger."

"So it kind of was this little pathway carved maybe three feet wide. And if you didn't make it, you'd hit a cliff and bounce down and basically die."

When the day arrived, the weather didn't cooperate. Dark skies and lashing rain threatened the jump. They had a small window in which to do it and the crew was ready to call it off. As he was taking off his gear, the skies parted and suddenly they were ready to go. Except now Jeb was filled with doubt.

"You have to remember, I had just hit Table Mountain. I had just recovered from a year of rehab, learning to walk again. And all of a sudden, here I am facing a project and facing something that's most likely going to put me right back in the hospital if I'm *lucky*. If I'm *lucky*, I'm going to the hospital. And I'm sitting there thinking, 'I'm not going to do this.'

"Everything I trained for my entire life was coming to this moment. And this was a transitional period of my life. This is a moment where I'm making a decision that's going to change the course of who I am as a human being. It was very overwhelming, because I realized I can decide not to do this right now and I probably won't die today. I'll probably live today. But in about 20 years from now, maybe 30, maybe 40, whatever the number is, I *will* die. And I'll look back at this moment as the moment that changed my life, where I finally allowed fear to dictate what I was going to do."

And all of a sudden, I think "You know what? Today's as good a day to die as any other. I'm going. And all of a sudden I became very calm, and I got on that helicopter and I jumped. And I started flying toward the cliff. And at first, I think to myself, 'Maybe I'll just fly over it ... I'm just going to fly over it. I'm not doing this.' And then I think, 'You didn't come here to fly over it. Just go for it.'"

And he did.

"It was one of the most powerful experiences of my entire life. And you can hear it in my voice. You're just hearing me scream, 'I made it. I'm alive. I'm alive.'

"Ten minutes before, this was done. It was never gonna happen. This was impossible. No way was this going down. And it was just amazing. It was one of those things where the experience was horrible, horrible, horrible. It's impossible to describe it. When you finally do step through fear and you find out what's on the other side of it.

"The fear at that moment, the moment of sitting in the helicopter contemplating the fact that I'd just recovered from an accident that should have killed me, and the fact that I survived it and then had to learn to walk again and went through multiple surgeries and the pain and the suffering.

"And I think the only reason I continued was because I knew I was going to die one day. That's really a big part of what's allowed me to do the things that I do. My little sister had a best friend who got her driver's license, had it for one week, was driving herself to school, got hit by a person who fell asleep behind the wheel, and killed her instantly. She'd never done a dangerous thing in her life. Boom, taken out doing something we do every single day without thinking about it. Another friend, brain aneurysm, walks into a room, dropped dead. No one even knew it. Healthy, in their 30s, didn't drink, didn't do drugs, nothing. Boom, turned off like a switch."

Jeb, like a lot of high sensation-seekers, doesn't have a death wish at all. What he has is an extreme passion for life and he wants to get the most out of it. For him, pushing to the other side of the crux makes him feel even more alive.

~

For each of these athletes it is apparent that their high sensation-seeking personality helps them to succeed. They have the ability to focus in a chaotic environment and the perseverance and grit to succeed, driven by the desire to crack through the crux and the amazing fortitude to bounce back when things don't work out. And while many people call high sensation-seekers like Will, Kristina, and Jeb adrenaline junkies it couldn't be further from the truth. Their thoughts about and physical reaction to adrenaline is fascinating. Amy noticed this when I talked to her about her time with the athletes.

"What was interesting was that their response to adrenaline depends on their years of experience. The most experienced people hated adrenaline, absolutely hated it and wished there was something they could do to get rid of it. The middle group, were in the zone and didn't hate it but tried to navigate around it and ignore it as much as possible. And the folks who were the newest to their sport were the ones who felt like it gets them going."

For many of these athletes there's another motivating factor, and it's another person in the room, sometimes thousands (and thousands) of others.

Doing it for the "Likes"

Facebook, Instagram, Pinterest, Snapchat, and Twitter have over 100,000,000 registered users.[35] In a few seconds you can grab your phone, send out a message and have it reach millions of people. In the past high sensation-seekers may have adventured alone, or with a group of people. Maybe they took some photos or a video, or just retold the story. But now it's just as easy to share your experience with the world, even as it happens.

Dr. Ann Pegoraro is a professor in the school of human kinetics and the director of the Institute for Sports Management at Laurentian University in Sudbury, Ontario. She teaches courses in Sport Marketing, Media and Sport, and Sport Communication. She lives, breathes and researches the digital world. And Twitter is her favorite. "I'm a bit of a twitterholic," she admitted. She's also a scientist that studies social media and its influence on athletes.

She wasn't one of the first people on Facebook but she quickly became fascinated when Twitter started to grow because she could see how quickly content moved and how people could interact in a new way. With her background in sport marketing she saw clearly how this tool could help fans and athletes connect to each other. She became fascinated by what athletes would do if they had unmediated access to people to promote themselves.

"It's a power that's fundamentally changed how we communicate because it's not always face to face. If I follow an athlete and interact with them, I sometimes think I know them better than I do."

I asked her about the influence of social media on sensation-seeking. "I do think that it does play a role because of how handy it is, how easy it is to share what you've done on your phone, on the phone that most people carry around, that you have access to videos or their own accounts where they're sharing the crazy things that they do.

"People doing that are looking for the followers to follow them and they are getting a high from that." But there's another thing. Social media can be really rewarding.

Mauricio Delgado, associate professor of psychology at Rutgers University in Newark, New Jersey, says that social media can have an effect similar to face-to-face interactions. "The same brain areas that are activated for food and water are activated for social stimuli," he says. "This can be a smile, someone telling you you're doing a great job or you're trustworthy, or you're a nice person, or even merely cooperating with somebody. All of these social reinforcers are abstract but show similar activity in the reward centers of the brain. This suggests that, perhaps, if you're getting positive feedback in social media – 'likes' and shares and retweets – it's a positive 'reinforcer' of using social media, and one that allows you to a) get the positive effects of it, and b) return to it seeking out more social reinforcement."[36]

So not only can you do it for the experience of the moment, but also now it's easy to get rewards from thousands of people you may not even know. Remember Jeb? Last time I checked he had over 70,000 Instagram followers (roughly the population of Napa, California or Kalamazoo, Michigan) all commenting and "liking" the things he posts.[37]

"I think they trigger some of the same things and reactions you get from the thrill-seeking behavior," Dr. Pegoraro explained.

"If all of a sudden now I have the most viewed video on Instagram or the most liked picture, or I get the most comments from people or they keep telling me how awesome I am and I should do more, what's the next thing. I think that feeds into the part of my brain that already wants to do that sort of thrill-seeking activity. So I think it probably acts more as a prompt and a regular trigger that you wouldn't always get. Because then you'd be relying on people in your life to tell you 'hey you should try the next big, best, hardest thing' and now you have complete strangers in the numbers of thousands following you and telling you that.

"A lot of athletes in sort of amateur sports have sort of realized that this is a way for them to build a personal brand and get sponsorships and allows them to continually still be a bobsledder or a luge in the three years that nobody cares about them between the Olympics. So from that perspective I think they see it as a necessary evil."

Agency and Taking Control of Your Life

A sense of accomplishment is not the only benefit that high sensation-seekers get from their activities. According to Matthew Barlow and his team at Bangor University, individuals may seek out situations of chaos, stress, and danger to demonstrate or reassert their agency and emotional control.[38] Agency is your ability to act in the world, to make a choice and exert power to make it happen. To be an agent means you are taking charge. Without it, we fold and let whatever happens happen. Emotional regulation is your ability to have control over your feelings. Without emotional regulation, your emotions are in control of you: toddlers have very little emotional regulation; actors require quite a bit. If you are frightened, the fear does not necessarily have to be expressed or consuming. Those with high levels of emotional regulation can suppress their fear when they need that extra push.

While nearly all risk-taking behaviors require some emotional regulation and agency, there are some that require considerable regulation for days, if not weeks, on end. And some risk-takers are actually motivated by their desire to control those emotions. Some researchers believe that there are people who sensation-seek to practice their sense of agency and emotional regulation. They then transfer this sense of emotional regulation and agency back to other areas of their life. Consider what Matt said about his experience doing Tough Mudder. After he made that first jump he felt invincible. After making it multiple times it wasn't even that scary to him anymore. This reduction in fear is something he takes back with him into his daily life. He felt his experiences at Tough Mudder had changed him on a fundamental level, perhaps all the way down to his DNA.

Looking for control on the mountain (or the Tough Mudder course) can be especially attractive to those who might not have it in their everyday life or to those who have a vague sense of dread, anxiety, and lack of control in their lives. Being in a situation where your actions have a direct impact on you and where the fear you experience is direct and palpable might be a refreshing change of pace.

It might be hard to imagine that a person who feels they have little control over his or her life could put themselves in a situation of great danger like scaling a mountain. But time and time again the sensation-seekers I have interviewed told just this kind of story. Having to focus their energy on a singular task in the here and now was highly therapeutic for them. Interestingly, the focus they achieve and the benefits of the experience mirror the outcomes some get from meditation. In mindfulness meditation focusing your awareness is a key component. What's more, regular meditation is associated with lower stress hormones and increased self-esteem.[39]

It's not unusual for people to feel a free flowing sense of anxiety that's not linked to any specific thing. Many of us live under a constant barrage of sensory stimulation: honking horns, loud noises, random people yelling (sometimes at you, and for no apparent reason). This can leave some people in a constant state of ambiguous anxiety. Swapping all that ambiguous anxiety out for a specific external fear that you can control may be therapeutic. Anxiety is mushy. Fear is targeted toward a known, specific threat, and there's more than enough fear in most risky activities. An externalized and directed fear can be diminished if the specific threat is controlled. In some cases, the emotion must be wrestled back to be able to achieve the task. This activity, creating and conquering fear, can have an impact beyond the mountain.

Not all risky activities involve the short-term "crux" experiences described earlier. Longer duration activities like ocean rowing or polar expeditions involve different kinds of emotional regulation and agency. Skydivers need to use these strengths only for a few seconds, while mountain climbers need them for hours at a time. In a way, the reward of their endurance activities is control of their sensations. There is some evidence that this emotional regulation lasts beyond the activity.[40] People learn from these experiences and use those coping skills in their everyday lives. Matt, for example, not only reports that facing his fears has made him less anxious in life but also that the cooperative nature of Tough Mudder helps him to be a better team player in the workplace. While these activities may not seem relaxing or even beneficial to some of us, for the high sensation-seeker there is clearly a reward.

Sports and adventures maybe be one of the most obvious ways that high sensation-seeking shows itself. It's easy to spot a high sensation-seeker in a bullring, bobsled, or the Beast Race. The way that high sensation-seekers process outside stimulation makes it easier for them to focus in such environments. But high sensation-seeking doesn't just affect them during extreme activities, it also pops up in their personal life too.

5 WHAT ABOUT YOUR FRIENDS: THE RELATIONSHIPS OF HIGH SENSATION-SEEKERS

Cindy was pretty excited when the guy she'd been dating for the last few months offered to fly her to Texas to visit her relatives. After all, she'd never been in a single-engine airplane before. The trip from Illinois to Texas would take a while. Single-engine airplanes fly considerably slower than commercial ones. Instead of zooming at 540 miles per hour you putter along at around 80, albeit in a mostly straight line.

During the long flight she peppered Tom with questions. "You know, the way you do when you're getting to know somebody and learning about what they're interested in," she explained.

Hours later Tom tapped her on the shoulder. Cindy was surprised. "I figured we were nearly there, except I looked down, and all I saw was water. I didn't know what we were doing out over open water. He had pulled us out over the Gulf of Mexico and lowered our altitude a bit. 'We're clearly not going to land there,' I thought." Then Tom does the unthinkable, he stalls the plane – deliberately. "We're in a single-engine plane over the ocean and he stalls it!" she explained with panic in her voice. "He said, 'All you have to do is take the plane out of the stall.' I screeched back, 'I can't. I've never flown a plane!' I was crazed, hysterical. We're in a tiny, tiny plane. Just a two-passenger stalled over the ocean!"

Tom explained exactly what to do: which knob to pull, which lever to move, and guided her hands every step of the way. Cindy had no idea what she was doing.

> "I pulled it out of the stall with his help. My first experience of being in a small plane was having to pull it out of a stall over the ocean . . . " she trailed off.
> "What was his reaction?" I asked.
> "Oh, he laughed."

Sensation-seekers like Tom live out particular patterns in their relationships with others. Their ways of dating, marriage, love, sex, friendship, conversation, humor, empathy, childrearing, even self-understanding and self-regulation, all bear the mark of sensation-seeking. In some cases, sensation-seeking helps people to deepen their experiences and relationships, and relationships with low sensation-seekers can help high sensation-seekers stay grounded and make better judgments about risks. In other cases, though, high sensation-seeking can detract from their relationships by blinding them to their loved ones' comfort zones.

By now you know the profile of a high sensation-seeker. They are on a quest for thrill and experiences but are easily bored and can be impulsive. It affects every aspect of their lives, from the way they drive to the music they listen to and how much Sriracha they squirt on their pizza. So it should come as no surprise that sensation-seeking also influences personal relationships. Sensation-seekers are drawn to things they find interesting and different and that influences the people they choose to be around and the way they interact with them. Sometimes sensation-seeking enhances the relationship, and sometimes it creates problems, not only for the sensation-seeker and their partner, but also for friends and families who are often along for quite a ride.

Interacting Differently

I fly a lot for work. As a low sensation-seeker, it's nearly impossible for me to get bored. I find my seat quickly, put away my bags, turn on my low sensation-seeking secret weapon – a noise-canceling headset – and I'm in low sensation-seeking bliss. For a high sensation-seeker, an airplane can be a source of excruciating tedium. There's not much to do and you are encouraged to stay in your seat with your seat belt buckled and tray table in its full upright and locked position, just waiting to land. You can't dance, or pace much, it's difficult to scratch your back; you can hardly squirm. In the absence of electronic entertainment, the only potential for experiences, the only source of escape, is roughly one inch away from you – the passenger beside you. After all, they could be an astronaut, or a rodeo clown, or maybe they'd love to hear your story about the time you saw a snake in the road, or your ideas about single-stream recycling, or give you advice about your boss.

Flying in an airplane is only slightly more stimulating than Zuckerman's original Ganzfeld sensory deprivation experiments (and you know how high sensation-seekers loved those). After those experiments Zuckerman began to wonder how high or low sensation-seeking people might respond when they were with another person either similar or different from them in sensation-seeking.[1] To determine this, he repeated the sensory deprivation experiments but this time with roommates. The experimenters were curious to what extent social isolation was a factor as opposed to the absence of sensory stimulation. When they paired high and low sensation-seeking people in rooms together they discovered that the low sensation-seekers preferred to be alone. Not only because the high sensation-seeking individual was a source of additional sensations, but also because the high sensation-seeker tended to dominate the social situation by talking more, disclosing more private information, and trying to convince the low sensation-seeker to disclose private information or even push hot button topics.[2] Zuckerman's follow-up study was one of the first early indicators that high sensation-seeking can impact a person's day-to-day interactions with others. They stand closer when they talk, make longer eye contact, and are more emotionally expressive during conversations.[3] For the high sensation-seeker, other people are a source of novel experiences. And they engage in higher rates of self-disclosure and initiate more disclosure from other people, including strangers.

In 1990, Robert Franken and his colleagues at the University of Calgary measured sensation-seeking in 413 people (158 men and 255 women) and asked them how likely they were to discuss personal information with casual and close friends. The researchers discovered that high sensation-seekers disclose both to casual and to close friends at a higher rate than low sensation-seekers. High sensation-seeking women in particular found it easy to disclose with close and casual friends. Both high sensation-seeking men and women indicated that they would be more likely to discuss sexual fantasies, attitudes, and anxieties with both close and casual friends than low sensation-seekers. What's more, high sensation-seekers had a higher tendency to encourage disclosure from others.[4]

High sensation-seekers sometimes have trouble defusing situations that might lead to altercations. By doing so they may be more likely to be pulled into high intensity situations (which they find stimulating). This can apply to their day-to-day interactions with others too. Getting people to disclose private details of their lives, and even telling risqué jokes are techniques that sensation-seekers employ to make their interactions more stimulating.

Sensation-seekers generally react to threatening situations with minimal amounts of negative feelings and dread. In fact they often see these threatening situations as a challenge or an adventure.[5] High sensation-seekers see people who are different from them as potential sources of excitement.

Low sensation-seekers prefer to be around people with whom they may be more similar, and when they are around people who are different, they may feel awkward and anxious. But there may be a way that some high sensation-seekers stack the deck in terms of finding these exciting differences. They poke the bear.

When given the opportunity to pick topics for discussion, HSSs are more likely to pick a topic of disagreement than average or low sensation-seekers.[6] Billy Thornton and his team at the University of Maine in Orono asked 135 undergraduates to complete a questionnaire.[7] They asked them to share their personal views and attitudes about subjects including God, war, drug laws, and premarital sex. They also used Zuckerman's sensation-seeking scale to measure their level of sensation-seeking.

After the attitude assessment, the students were told that later in the day they would meet with another person to have a discussion and they were presented with the attitude survey of that person. In actuality this was a bogus attitude survey that was selected to either be very similar to the person (60 percent agreement in attitudes) or very dissimilar (only 30 percent agreement in attitudes). The undergraduates were asked to look over the attitude survey and then rate how much they were looking forward to meeting the person, how much they expected to like the person, and here's the fun part, to pick a topic of discussion from the survey of attitude items. Low sensation-seekers were more attracted to the people most similar to them and chose a topic they were likely to agree on. The high sensation-seekers were more attracted to dissimilar others. What's more, high sensation-seekers were more likely to pick topics they knew would lead to more novelty and challenge. They picked the things that were more likely to lead to disagreement.

Lynn wouldn't be surprised by this finding. Lynn is a 51-year-old business owner who told me about her friend Todd. "Todd is amazing. If he thinks he can get somebody to do something, it becomes a game for him to get that person to do that thing. Whether it's a physical thing or a business deal or something like that." Todd and her husband used to be close. But they stopped hanging out in college because Todd was doing things that were too risky.

Thrill-seeking was the core of her husband's friendship with Todd. "Todd was one of those kids that pushed everyone's buttons in school. Todd was a really fast runner, which was good because the kids were always trying to chase him to beat up on him. He would create controversial situations. He absolutely knows how to push people's buttons. That's part of his game, part of his thrill."

That's What Friends are For?

"Keep smiling, keep shining, knowing you can always count on me, for sure. That's what friends are for," or so the song goes. But psychologists have a different (colder and more scientific) view of what friendship does and how we operate in the social world. Psychologist Susan Fiske believes that friendships, well, nearly all social connections, are motivated by just a handful of aims.

She calls it the BUCKET theory because the words (nearly) spell out the word "bucket." *Belonging*: people are motivated to join together. They want to be part of something. *Understanding*: people are motivated to make sense of the world. *Controlling*: people are motivated to have power over the things that happen to them. *Enhancing self*: people are motivated to feel and be perceived as worthy. *Trusting*: people are motivated to see the world in a positive light.

What kinds of people do high sensation-seekers typically seek out for friendship? And how does being an HSS impact the ability to attract friends and the quality of the friendships themselves?

HSSs are really no different than the rest of us when it comes to their motivations, except of course some of their motivations have to do with increasing sensations and having a high tolerance for activities that create sensation.

I talked to Ava, an HSS public relations specialist, about her friendships. Ava loves to be in the air. "I love doing new things. I'm constantly trying to go on adventures and do new things. I like scuba diving, skydiving, I like exploring the air and the water. I like hiking, roller coasters, bungee jumping, or parasailing. Oh, and zip lining. I love, love, LOVE zip lining. Anything that's somehow like flying." Ava's been a thrill-seeker since she was a kid. "By fourth grade," she recalls, "I was wearing 'No Fear' shirts every single day. That was the beginning of me having this no fear attitude and just doing whatever."

Researchers have found that the higher your total sensation-seeking score, the more friends you are likely to have.[8] This doesn't mean that average or low sensation-seekers have trouble making friends or make bad friends, it may just mean that sensation-seekers desire more friends to help them do the things they love doing or just to get to know. Clearly, sensation-seekers are more likely to bump into other sensation-seekers involved in similar activities. So, you'd imagine that high sensation-seekers have lots of HSS friends. What's more, remember how HSSs love to disclose information? Self-disclosure of personal details is an important aspect of friendship.[9]

Lynn says "As friends go, the thrill-seekers find each other. Especially the ones that want to do things that are on the edge."

In some cases, this may be true. But this isn't universally the case. Research has shown that sensation-seekers are drawn to people who do the things they love doing, but sensation-seekers find all sorts of things about people interesting, so they are also attracted to people who are dissimilar to them.[10]

This might be because learning about someone different from them is part of the thrill. But it also means that some HSSs feel misunderstood by their friends.

Ava's experience bears this out. "I feel like I have a lot of friends, and I have a lot of different kinds of friends. But usually when I'm doing something that's a bit scary, it's solo. I'm constantly inviting people to do things with me." They usually don't. Why not? Most of her friends aren't thrill-seekers. Her response to them? "*Mais fica*," which is Portuguese for "more for me."

Because of her "no fear" lifestyle some of Ava's friends don't exactly respond positively to some of her choices of activities.

"Oh here goes Ava doing her wild stuff again," she explains. "At this point it's almost like I don't care how they feel anymore. Sometimes I'll invite people. But sometimes I won't, because it's almost not even worth it anymore. And sometimes I'll just tell them once I'm back," she said in an exasperated tone. "In October, I went to Egypt and Dubai for 17 days, and I didn't tell a lot of people just because I didn't want to hear it. I didn't want to hear why I shouldn't go or why it's dangerous, or how I should be safe, or whatever tips they think they are going to try to give me."

A lot of sensation-seekers feel misunderstood as being "crazy" or "wild." People wonder what's wrong with them because of their sensation-seeking activities. Having friends who understand and support their unique experience of the world can be an important aspect to their friendships and give them a sense of belonging (the "Understanding" and "Belonging" in the Bucket Theory). People who affirm this social identity are more likely to share the intimate parts of their personality. It doesn't have to be other high sensation-seekers, just another person who supports the things they do.

A high sensation-seeker's ability to tolerate intense experiences goes beyond vertical spelunking. It goes into interpersonal interactions. There are some social interactions that can be intense. Some average and low sensation-seekers will just avoid those situations altogether, just to make the interaction smoother.

Let's say that you don't seem to react to those potentially stressful situations in a typically stressful way, and let's say that you tend to be less inhibited. That could be a recipe for a "say whatever you want" type of attitude. Ava says she gets described as "intense" and it runs in the family.

"My sister and I are both pretty intense, I think we are pretty *real* as far as the conversations we like to have and what we might say to someone who's our friend. We're not going to sugarcoat it that much." She continued, "I went to my friend's wedding in North Carolina and it's all these girls telling the bride that her hair looks great. The lady had put this giant flower in it and it looked awful. I told her, 'You know what? That flower is not working,' and she was like, 'Oh thank you. I didn't like it but I didn't know what to do.'"

As previously mentioned, research suggests that thrill-seekers can process intense emotional experiences in a different way. Situations like this just don't feel as overwhelming to them, because they can think through the experience in a clearheaded way. This can carry over into relationships as well. Where another person might have felt frightened to confront the bride, Ava and other high sensation-seekers might not even think twice about it. Like Ava says, "No fear."

This combination of reactions makes it easier for Ava to control interpersonal situations around her and it also makes her a natural leader when she's in a group. "I'm definitely always the leader. If there's a group of people walking I'm somehow always in the front, I'm always leading it. Even if I don't know where we are going. If I'm on an airplane I want to be in the exit row. I'm the one that says 'Okay everyone I'll open the door, let's go down the slide, it's going to be okay.' I don't want someone who's panicking. I know that I'm going to be able to handle it" (Controlling: the "C" in Fiske's Bucket Theory).

So where does all of this leave the sensation-seeker in friendships? For the HSS, friends provide a source of entertainment, encouragement, and support for their thrill-seeking identity. This isn't so different than the rest of us. However, research has shown there are some distinctions between the high sensation-seeker and the low sensation-seeker when it comes to building friendships as we've outlined earlier. We also know that sensation-seekers are more likely to try to get others to divulge juicy details of their lives. This means they may be less likely to be drawn to people who hold back interpersonally or take a lot of time to warm up to new people. However, it also means that it's relatively easy for them to get closer to others. Because they are good at getting people to disclose personal information and they are relatively open communicators themselves, the HSS may end up in close friendships more easily.

Intercultural Friendships

Talking to people who are different than you can be a challenge. It can be easy to be misunderstood. It's even harder when the cultural norms are different. My mother was big on manners – which fork to use, always saying please and thank you, and being polite to others. It was reassuring to be certain you wouldn't insult other people and knowing exactly what to say. Fast forward to my first time at summer camp. Each week we had the option to attend a different religious service. I've never been so confused. When to stand, when to sit, what and when to sing – and that was in my own city!

When it comes to intercultural communication and etiquette things can be tricky. So much so that nearly everyone who has traveled internationally has had an experience where they were insulted, or have insulted someone completely unintentionally, or were just plain baffled. The hotel group Swiss Hotel has an online guide to help. They call it "The Ultimate Guide to Worldwide Etiquette."[11] You pick a country, and the website gives you information on tipping, gestures, dining, and do's and don'ts. What it reveals is that there are lots of ways to mess up and there are etiquette landmines everywhere. In the United States it's just fine to fill your own glass, not so in Turkey, China, and Egypt where it's more polite to fill the glasses of others. Finished your meal in Cuba or Italy? Simply place your fork and knife on the right side of the plate to let the server know you are done. Put them *on* your plate in Spain and Australia, but in South Africa that means you are still eating. In Switzerland, tapping on your temple means "crazy" and it means "clever" in the Netherlands (big difference). What's more, these customs can vary from situation to situation and from formal to informal interactions.

With all of this going on it might be difficult to build friendships and have positive non-blundering interactions with the people you meet. I can understand how interacting with people across cultures can create anxiety or step on toes.

It turns out that high sensation-seekers are really good at just this kind of situation and they are more likely to seek out intercultural interactions than low sensation-seekers. Why? Intercultural encounters can be novel and ambiguous in nature and HSSs are most interested in connecting with people who are at least moderately different from them.[12] Sensation-seeking predicts intercultural contact seeking. A positive attitude toward people of other cultures leads to motivation to communicate with them which leads to more experience. Intercultural communications competence is related to high sensation-seeking in both the effectiveness and the appropriateness of what the high sensation-seekers do. Because high sensation-seekers have more interest and empathy in people who are different from them, it means that the high sensation-seekers listen more carefully, which increases intercultural competency, especially in unfamiliar settings.[13]

Intercultural communication can require a person to master nuances and manage the anxiety of not being sure what to do. There are unfamiliar norms and shifting expectations. The process requires empathy, patience, and tolerance for ambiguity, and knowing what we know about sensation-seekers, it seems they are up for this challenge.[14]

HSSs find intercultural interaction to be more fun and satisfying than LSSs and they have attitudes and behaviors that make these friendships more successful for them. The increased comfort leads to positive attitudes, and more experience and increased competency. And all of this leads to better intercultural communication. In a study of intercultural friendships, Susan Morgan and Lili Arasaratnam found that high sensation-seekers were more likely to indicate that they find it "boring" to always associate with people of their own culture and that they "enjoy" initiating conversations with someone from a different culture.[15]

For the high sensation-seeker, intercultural communication is also a form of social risk-taking and HSSs have the biological predisposition to form friendships in risky social environments. Some high sensation-seekers are driven by the mystery and curiosity of these experiences.[16] The high levels of uncertainty make this a particularly appealing challenge for a high sensation-seeker.

Loving Differently

It's easy to see how the outgoing, engaging style of sensation-seekers could be very beneficial in certain circumstances. There are situations in which standing closer, talking louder, and making more eye contact can be extremely useful – for example, when meeting new people in a bar or nightclub.

You see, high sensation-seekers are attracted to a more ludic style of love. To understand what that means, and to better comprehend what love is and how it works, we need to take a look at the work of sociologist John Allen Lee in the 1970s.

The Six Styles of Love

In 1973, Lee interviewed hundreds of people in order to find out more about love.[17] He was determined to try to explain and categorize the different ways that people love. Through his interviews he created a taxonomy, a way to tag various kinds of love, and gave each one a Greek name: *agape, eros, storge, pragma, mania*, and *ludos*.

Agape is selfless, non-demanding, altruistic love. Full of commitment, *agape* is unconditional and unbreakable. *Eros* is romantic, passionate love, the stuff of novels. Erotic love is fast and deep, emotional and intense. It's based on physical attraction and strong commitment. *Storge* is friendship love. In storgic love, there are lower levels of passion, but deep respect and commitment are key. It's a merging of friendship and love. Physical attraction is less important. Shared activities form the basis of the bond. While there aren't intense emotional or physical connections, the shared interest and commitment lead to an enduring connection. *Pragma* is practical, logical love. The pragmatic lover has already decided who she or he should love and has a ready list of attributes the person must have. It's deep in commitment but shallow in passion. They are looking for a compatible mate to share common goals. *Mania* is possessive and dependent love. Often the manic lover hopes to win the love of another. It's an unhappy state to be in. They want desperately to be loved and may even try to win the love of others. This type of love is often jealous and even more often unhappy.

Finally, *ludos* is game-playing love. The hunt for love and attention is all part of the game. They love the chase. Emotional involvement isn't as deep, and ludic lovers may even be suspicious of commitment. Sex is for pleasure and not a bonding experience. Ludic love can involve deception, manipulation, and multiple partners.

Of these six types of love, sensation-seekers tend to score highest in ludos, the game-playing love.[18] This attraction to ludic love may be why long-term relationships can be problematic for some sensation-seekers. Sensation-seekers don't always look at relationships for their long-term potential. The higher the sensation-seeking score, the less likely they are drawn to commitment in relationships. This may be especially true for those with high levels of disinhibition and boredom susceptibility.

If you think about it, being in a successful long-term relationship means being successful with two very different types of skills. Being good at finding someone means you need to be flirty, open to meeting lots of people, and into the thrill of the hunt. Being in a long-term relationship calls for commitment, compromise, and a willingness to be vulnerable to letting someone know you deeply, but it also requires a certain boredom tolerance – something some sensation-seekers just aren't good at. Being with the same person year after year just doesn't provide the same thrill that hunting for a new relationship delivers.

Some sensation-seekers find they are pretty good at the first part but may struggle with the second. The thrill of the hunt for a new relationship might not deliver the same excitement as being with the same person year after year. This doesn't mean that they are doomed to a life of solitude or serial monogamy. It might mean that some sensation-seekers with high boredom susceptibility who want a long-term relationship may need to find the shared excitement and discovery only people in long-term committed relationships can have together.

High sensation-seekers may be more likely to keep their options open to have as many of those new and interesting activities available to them as possible.[19] A total of 986 participants filled out Zuckerman's sensation-seeking survey and also filled out Robert Franken's "Keeping Your Options Open Scale," which measured a tendency to put off making commitments, act impulsively, and break commitments if something more interesting comes up. Not surprisingly, sensation-seekers have a general disposition to keep their options open and were more likely to indicate that they "like to maintain the option of being able to do what I want when I want even if people think I'm undependable." They may do this for relationships as well.

Remember Anne, the woman in Chapter 3 who dropped everything, gave up her apartment, and took off for Samoa on a whim? She told me that she loves skydiving, zip-lining, and flying in helicopters (preferably over active volcanos). She chuckled when I asked her about jumping out of perfectly good planes. "My main opponent in everyday life is boredom, and high sensation-seekers do not only jump out of perfectly good planes," she told me. "I also leave perfectly good jobs, homes, and relationships, because I get bored. It definitely happens in my sex life. I do seem to get bored with the same partners after a period of time, no matter how exciting it is. Two years is usually the cut off. Now, I don't want to give you the wrong idea. That doesn't mean that I cheat on anyone. It's more like I'm a serial monogamist. If that makes sense."

It's probably obvious that sensation-seeking would predict certain things about an individual's sex life. Those with high sensation-seeking personalities seek greater amounts and more intense sensations and arousal in the bedroom. Sensation-seekers have more permissive attitudes toward sex and more varied types of sexual experiences with more partners.[20] Sensation-seeking individuals may also seek more intense sexual situations and unplanned encounters.[21]

Rick Hoyle and his colleagues Michele Fejfer and Joshua Miller reviewed all the studies relating to major personality traits and sexual risk factors.[22] Sexual risk-taking was defined by number of partners, having unprotected sex, and high-risk sexual encounters such as having sex with a stranger. Sensation-seeking is correlated with all three. In fact, when you take into account all types of personality traits, sensation-seeking may be the best way to predict risky sexual encounters.

For example, in the young adult population, high sensation-seekers were more likely to have had sex, intended to have sex in the near future, had unwanted sex under pressure, and were more likely to have had unwanted sex when drunk.

However, it isn't only their attraction to ludic love that may make it difficult for some HSSs to maintain long-term relationships. It turns out that, for all their gifts, some high sensation-seekers may struggle with a core concept in relating with other people – emotional intelligence.

The Challenge of Emotional Intelligence

Emotional intelligence is your ability to perceive and regulate emotions. It helps us to be socially skilled. Peter Salovey and John Mayer coined the term in the 1990s to convey "a form of social intelligence that involves the ability to monitor one's own and other's feelings and emotions, to discriminate among them, and to use this information to guide one's thinking and action." In one study people who were good at perceiving emotions accurately were able to respond more quickly to shifts in social environments like knowing when they've taken a joke too far.[23] Later, Daniel Goleman built on the work of Salovey and Mayer. He suggested that cognitive intelligence alone wasn't enough for success but that emotional intelligence was important as well.[24]

There are two important components of emotional intelligence: intrapersonal and interpersonal intelligence. Let's start with the first one. Intrapersonal intelligence is the ability to understand your own emotions and to be able to alter those emotions. Knowing when you are feeling grumpy is one thing. But being able to flip off the grumpiness to seem chipper when you are actually really annoyed is the stuff of intrapersonal wizardry.

Early in grad school I took a trip to Toronto for a psychology conference. I scored an amazing hotel room for a steal. The hotel was tucked away on two floors of a bank building. They served breakfast, snacks, and hot foods in the afternoon. I was ready to move in.

I asked a person at the front desk if I could grab a few breakfast muffins and take them to my room. He got this look of delight on his face and said "absolutely." His facial expression conveyed that not only could I do it, but he would also like to congratulate me for even coming up with such a wonderful idea. I even felt proud to have thought of it. By the time I got back to my room, I noticed others taking muffins back to their room. Surely they hadn't overheard my brilliant plan. Then it hit me. I bet he gets asked that question 20 times a day. His ability to conjure up that emotion in himself is intrapersonal intelligence, but his ability to summon an emotion in me shows he's a pro at *inter*personal intelligence as well.

Interpersonal intelligence is the ability to notice the emotions in others as well as the ability to alter emotions in them. Used together, intrapersonal and interpersonal intelligence constitutes our overall emotional intelligence – which can be a very powerful force. People with strong emotional intelligence are experts at knowing how others might be feeling, and they are pretty good at directing others to feel a certain way. Without emotional intelligence, you are cut off from others and sometimes unaware that you are.

But what does this have to do with sensation-seeking? A study of sensation-seekers found that they tend to be lower in emotional intelligence.[25] Their interpersonal intelligence seems to be particularly impacted. This may be because they are so distracted by their own sensations, they can't tell if you are frightened or annoyed by a particular situation. Remember Cindy, whose new beau stalled the plane over the Gulf of Mexico? He may not have realized how terrified Cindy was at the time. Not only was he low in altitude but he was also low in emotional intelligence. "He didn't feel what I felt," she explained to me. "He understood it, but he didn't feel it. When I brought it up, he said, 'That's silly. Don't be afraid. The plane won't actually crash. It just won't happen.'" Easy for him to say.

Their difficulty with interpersonal intelligence may be yet another reason some high sensation-seekers struggle with long-term relationships and have higher divorce rates and lower relationship satisfaction than average or low sensation-seekers.[26]

It may sound as though I'm painting a grim picture of the high sensation-seeking person in romantic relationships. Some sensation-seekers may just see love and sex differently than the rest of us do. It might also mean that couples who are interested in a long-term commitment could benefit from finding shared interests that can lead to excitement and discovery for both people. This may be particularly important for high sensation-seekers with high boredom susceptibility.

At first the idea that high sensation-seekers may have struggles with emotional intelligence seems counter to the evidence that they have more empathy in intercultural communications, but that's just it. They seem to only have higher levels of empathy with people who are culturally different from them. It seems that the more culturally similar you are to a high sensation-seeker, the less likely they may focus on what your experience might be.

Intense Experiences Leads to Intense Love

While some high sensation-seekers have a proclivity to promiscuity, this doesn't mean they all remain single or in serial monogamous relationships. I've met many high sensation-seekers who successfully maintain deep and meaningful long-term relationships. Take Kris and Jess, a couple of sensation-seekers in their early thirties. They are academic linguists who engage in what they call "outdoor shenanigans" when they aren't researching, writing, or teaching classes. These shenanigans include alpine climbing, sport climbing (with ropes), and bouldering. Bouldering involves scaling along a house-sized rock with no ropes, but there are pads underneath to soften the fall.

"Fall?" I asked.

"Oh, yeah, there's lots of falling involved," Jess offered emphatically.

Jess and Kris have been married for over ten years and they couldn't be happier or more attuned to each other, in fact they often finish each other's sentences. They were both interested in climbing, skiing, and hiking since long before they met. Their first date? Twenty-four hours of hiking. Weeks later it was a ten-day backpacking trek.

"It's been ten days and you're with this person you just started dating with no showers, no civilization, carrying a lot of crap," Jess said. "And you only have each other to talk to."

"Wow. That's love," I said blissfully.

"That's something," they said in unison.

It's hard to imagine being on a cliff watching your loved one struggle up a rock wall and in danger of falling. I'd be so terrified that it would be nearly impossible for me to do anything but cross my fingers and whisper prayers the entire time. I asked them if they ever get frightened for each other when on these adventures.

"Typically I'm not overwhelmingly scared for him or myself," Jess said. "I trust in his ability to look at the risk and my ability to look at the risk."

"I'm not worried when she's climbing." Kris continued, "I'm generally not worried for her because I know that she isn't going to push it to the point where it's dangerous for her. If something goes wrong, my mind turns immediately to problem solving. When I'm in a situation where it's sketchy my mind pushes that fear away and says, 'How do I solve this problem?' That's the way my mind operates, I guess."

The same ability (or inability) to perceive risk, the same trust in their ability to cope with stressful situations that applies to individuals seems to apply to couples as well. In a way, HSS couples are in their own bubble. And that bubble seems to enhance their relationship.

"I think sometimes our climbing has probably added a different dynamic to our relationship, because we really, really, really have to trust each other," Jess explained. "If we're in a situation and I say, 'Kris, I don't think this is safe,' and he says 'It is. We need to do this to get out of here', I have to say, 'Okay' and I have to turn my brain off and just trust that."

It's true. High sensation-seeking couples find themselves in the middle of intense experiences where their decisions are really important: sometimes life or death. Think about it. A lot of couples will say, "I trust my partner with my life," but how many couples can actually point to a specific (or dozens of specific) examples where that was true? "It probably makes fighting about the dishwasher pale in comparison," I said.

"I don't know about that," Jess rolled her eyes.

It's always the dishwasher.

But the trust that develops in HSS relationships doesn't only help strengthen the trust in their everyday lives, it enhances the sensation-seeking activity too.

"I feel like our climbing experience and our experiences outside of climbing are made fuller because we get to experience it with each other," Jess said. "Because we have a level of trust outside of our activities that most people who just have a random climbing partner don't really have. I think the relationship enriches the climbing experiences and vice versa."

But couples who spend their time engaged in such shenanigans feel just as misunderstood as people who do it alone. Parents and friends who are average or low sensation-seekers sometimes tolerate it, but don't necessarily get why they do it.

"I don't think your parents really understand it," Jess started. "My dad think's it's awesome, but I don't think he understands it really."

"I would say they're supportive of us," Kris added. "In the same way you just sort of accept the crazy aunt or uncle for who they are, they accept that we are going to climb and it's not going to change. Neither one of us inherited any of our activities or anything from our parents."

I asked if some of their activities might change if they had kids.

"Yeah, you know, we've talked about it some actually," Kris explained. "I think that instances where we are both in a situation that if something went really wrong, both people would, you know . . . "

"Die," Jess finished.

"Yeah, if both people were in danger of dying, we've talked about perhaps not doing that so much, or perhaps not even at all."

Perhaps.

"But the actual climbing lifestyle? I don't think it's going to change," Kris said.

"No. We would just put them in a baby carrier and strap them to a tree," Jess joked. (I think.)

"Right," Kris agreed in sort of a verbal high five.

Greater trust, shared wonderful experiences? Sounds fantastic. But all this made me wonder if a high sensation-seeker could get along just as well with someone who doesn't share this personality trait. Jess wasn't so sure.

"If they have the draw toward some kind of adventurous activity and they start dating somebody who doesn't, that's definitely something that they should probably figure out very early on," she cautioned. "Kris and I both had other serious relationships. My most serious relationship before him, the person was not very adventurous. I just felt a constant pulling, like I was having to drag someone along. It was kind of exhausting because I felt like I was always trying to encourage him, get him going. I don't have to do that with Kris. For me I can't imagine being with somebody who can't enjoy the same kind of experiences that give me complete fulfillment."

Does that mean that the only way a high sensation-seeker can be happy is if they find a like-minded high sensation-seeking partner? Not necessarily. Sometimes an HSS and an LSS meet, fall in love, and make it work no matter how unlikely that may seem. Consider the story of Gina and Ed.

An Anchor that Keeps the Intensity at Bay

Gina is a high sensation-seeker who has known she was a little different since she was a kid. "It all started when I was about three or four. I would be outside playing with my older brother, and I would look up and see the trees or the top of our house and feel the need to climb. If my mom couldn't find me, she'd look on the roof because chances are, that's where I'd be." Her parents' pleading didn't seem to hold her back. "'Oh, get down, you're going to hurt yourself! You're going to break something! You're going to fall!' That never slowed me down. One time they had to call the fire department, because I refused to come down. I was having fun."

As she got older her interests shifted – rock climbing, rappelling, spelunking, white water rafting, polar plunging, and skydiving. The older she got the more challenging the thrills. Then at a church event, she met Ed. Ed's relatives happened to live right next door in her little town of about 800 people. Ed isn't like Gina; he's more like me, a chill-seeker rather than a thrill-seeker.

"He cheers me on from the sidelines. When I'm going skydiving, he says, 'I'll take pictures.'" I asked her if there's anything that she does that's a problem for him. Without hesitation, she said, "Driving. I tend to be a little aggressive as a driver. I've had a number of speeding tickets, because I'm in that 'go' mode and not paying attention to how fast I'm going. My husband refuses to be the passenger in the car when I'm driving. He's like, 'I need to drive. You can't drive.'"

Gina says that now her husband is her anchor. "He keeps me grounded, so I have a somewhat 'normal' life. He says things like, 'Remember you're my wife, you're the mother of our two kids. You can't go off and always do these crazy things. You could get yourself killed. Then the kids aren't going to have a mom.'"

"I have to remind myself that it's not just me anymore. When I was single, it was just me. But now I have to think about them too."

Even with their relatively low emotional intelligence, propensity toward ludic love, and higher divorce rates, plenty of HSSs make long-term relationships work. It may take a little additional effort and some special arrangements between partners, but it certainly does happen.

That said, it is clear that being a high sensation-seeker can impact intimate and romantic relationships. But those aren't the only relationships we humans engage in. Being a high sensation-seeker also influences friendships and families.

Would You Feed Your Kids Pufferfish?

We know that high sensation-seekers report having stimulating family environments and that high sensation-seeking tends to run in families. It's likely there is a genetic component to this, especially from a biochemical standpoint. It seems very probable that the children of sensation-seekers inherit their parents' reactivity to dopamine and relative lack of reactivity to cortisol – setting them up to seek out the same extreme behaviors their folks do. Gina, from the story above, clearly thought her children had inherited her high sensation-seeking tendencies. "I see it in my own two children, especially my youngest. Even as a toddler, he had no fear. At swim lessons, he would jump right into the 12-foot deep end. I had to tell him, 'You can't just do that! You're still learning how to swim.' But there's no fear there. He's nine now, and I don't know of anything he's afraid of."

Clearly nature plays a role, but nurture probably does too. The thrills sensation-seekers create in their family life could have something to do with their children's predilections. Remember Jimmy, the fearless foodie from Chapter 3 who tried the Japanese pufferfish? He, like lots of sensation-seekers, encourages his kid to try the new experiences that he engages in.

"I don't push the envelope on my kid's eating just for the sake of it," Jimmy told me. "I feed him what I'm having on my plate. If that's going to be a curry laksa then so be it. Two weeks ago we went for Indian food and he was eating things that have a decent amount of heat to them. I made some quesadillas the other morning with chorizo, potato, and eggs, kind of breakfast style. I was dipping mine into a real hot sauce. He saw that and wanted to do the same, so I poured him some. He was dipping his quesadilla into this Valentina hot sauce and enjoying it.

"If I'm confident that the food's been prepared properly and sourced properly, then I have no problem feeding him whatever I'm eating. I hope he likes food. I know little kids sometimes get pickier as they get older. We'll see."

I asked him about a funny video I'd seen of the expressions that kids make when trying lemons. I wondered if his son made faces like that, or if he had tried to give his son a lemon like they did in the videos. "Yeah, my wife and I were really excited to do that to him, because it's hilarious. And he just ate the lemon and loved it. He didn't make the face. Now when he sees lemons, he wants to suck on them." Sounds like a junior sensation-seeker in the making.

But these sensation-seekers in training aren't the only family members impacted by HSS-type behavior. The sensation-seekers themselves are often affected. Many sensation-seekers say that once they have children, their sensation-seeking declines. This was true of Gina, for example. We know that sensation-seeking decreases as people get older, but many sensation-seekers have told me that thinking of their family, friends, and children reduces their disinhibition and thrill-seeking behavior. They think twice about their sensation-seeking activities. It's not fear that motivates them to change their behavior, but what might happen to their families and how they would feel if things didn't go as smoothly as planned. It's not just partners but family and friends that act as anchors to keep sensation-seekers rooted and steady in some cases.

~

What I ultimately took away from all of this was that being a high sensation-seeker absolutely does impact relationships. But high sensation-seekers still have a wide variety of relationships (maybe more than most), and the way that their behaviors affect their friendships, intimate relationships, and families, changes on a case-by-case basis. Consider two of the couples featured in this chapter and how things worked out differently for each of them.

Despite the frightful date, Cindy and Tom continued to see each other and later married. But things didn't work out. Sensation-seeking Tom relied on Cindy to be his down-to-earth anchor – but ultimately Tom didn't really want an anchor. They divorced three years later.

Other couples, like Gina and her husband, find ways to combine the interests and comfort of differing levels of sensation-seeking by embarking on experience-seeking activities together.

"My husband and I have been to every winery in the state of Illinois. Over 110," Gina shared. "It was actually my husband's idea, because we got to looking at the map, and he was like, 'We could do this. It'll take us several years, but we could do this.' I was excited. It tapped into a new part of my experience-seeking side. Now we have a notebook with a different winery on every page. It shows the wines we sampled, and what we got out of it. The experience has been amazing, going to all these different places. It's something we enjoy together."

~

It's clear that high sensation-seeking has a profound impact on day-to-day relationships. From jokes to dating to picking topics of conversation, high sensation-seekers' personalities shape their connections with others. But your personality doesn't just kick in after 5pm. You carry it around with you everywhere you go. Even at work.

6 ALL IN A DAY'S WORK

Imagine plodding through a pool full of beef stew: thick, murky, and carrot-filled. Now add the rushing currents of a storm surge and you have an idea of black water. There's zero visibility and chunks of debris fly about. Extreme scuba diving in black water during the huge tidal surge on the east coast that came with Hurricane Sandy was not something even the most fearless high sensation-seeking people would have done. For Jason this isn't a hobby. It's his job. For the last 15 years, he has been braving conditions like these and worse to fulfill his life's passion and profession: to discover new species and better understand ancient life.

Jason's career is far from typical. He is a mechanical engineer, project strategist, paleontologist, and runs a non-profit. He's been on archeology digs. And then there is his diving, which facilitates the deep need behind all of these endeavors – an insatiable desire to discover new things.

"It started in childhood, and continues to the present. Throughout my life, this desire evolved from an interest to a passion, to an obsession, to a business, and now into this Indiana Jones style of field work." Which is exactly how I think of Jason – the underwater Indiana Jones who braves treacherous conditions for treasure. Jason spends a major portion of his life tying heavy weights to himself – weights so heavy that he has to climb (not swim) out after his dives – trolling along the bottom of river beds using only a bright light that allows him to see a mere 8 to 12 inches in front of his face and a screwdriver that he sticks into the clay and mud to help drag himself along. This is absolutely as dangerous as it sounds. Jason has found himself trapped inside an underwater cave, wondering how he got there and how he would escape. He's come face-to-face with 7-foot long crocodiles (for the record, he prefers to approach them from behind instead of head on). He's been physically bumped by bull sharks *and* he regularly gets tangled in debris like fishing line, nets, tree roots, and more, while weighted down under water.

Oh, and this is in the middle of hurricanes and other extreme weather raging around him. He dives during such tumult on purpose apparently because that's the best way to find things: "If we get really extreme conditions in rivers that are cutting through ancient formations, they start ripping out new material. You have to put yourself, put your body into that position in order to be successful [i.e. to find new things]. You want to be there right at the time that that happens. Most of the time I'm diving during storms or if there's flood conditions and black water and there's logs and everything else flowing down the rivers. I'm at the bottom of the river. I have to weight myself, make sure I stick to the bottom, so I don't get washed away. I crawl on the bottom and pull up new species or whale bones or fossils."

You wouldn't think it would be possible to have the poise to figure out what's old and what's new while under water in the middle of a raging storm, but Jason says he's calm during these dives. "There is a Zen factor of really putting your body into this situation. You're in a zone. It's unparalleled. Your whole world is 8 inches in front of your face. You're not thinking about anything else, because you can't. You have to think about that world. There could be bull sharks swimming around you, you won't see them. There could be alligators above you. You just don't know. You're just pushing yourself along with the anticipation that you're going to run into something, discover something that no one has seen before. It's really interesting."

Interesting seems like an understatement.

"After a while it becomes a comfort zone. It doesn't feel risky. It seems absolutely natural. It's kind of crazy." Yes, it does seem a bit crazy to us low sensation-seekers, but what I find fascinating is that Jason not only remains calm, even intensely focused, in the most extreme of circumstances, but he also actually craves it. It's almost like he *needs* to dive or it starts impacting his life in damaging ways.

"I crave it. I have to do it. If I don't do it every so often, I go nuts. Again, it's that anticipation – getting out there and being in that environment – it's pretty crazy."

As you know by now, being calm and even intensely focused in circumstances where most of us would crumble and panic is a hallmark of the high sensation-seeking person. In fact, there is likely a physiological component to this reaction as I described in Chapter 2. Here is my guess about what is likely going on within Jason. Cortisol and the other corticosteroids naturally released in stressful situations are probably available in smaller amounts than they would be in low sensation-seekers like me. This allows for the hyper-focused Zen-like state he refers to in a situation that would be utterly overwhelming for most. After all, the stress response can be a good thing in that it shuts down peripheral activities in the body and leads to a heightened awareness and greater attention to the present moment. In addition, Jason probably gets a big boost of dopamine when he does these dives, which makes the experience rewarding for him. That's probably why he can't go for long without doing a black water dive before he starts to feel like he's "going nuts." This is probably also why he craves the experience. We all crave activities we know will be rewarding, and we know that cravings (even extreme forms like addiction) are connected to the dopamine response. So his behavioral approach system is hard at work, while his fight or flight response is in an optimal state of arousal. It's pretty clear that this magical combination – balanced stress and increased reward – provides the neurochemical environment necessary to make black water diving a crave-worthy experience for him, whereas it would feel like a chaotic murky hell to someone like me.

There's no question that Jason is particularly extreme, but what we see here is what I heard over and over again in talking to high sensation-seekers: a deep desire to do something out of the ordinary, a Zen-like sense of balance in the face of amazing circumstances, and the desire, the *need*, to repeat this over and over again. It's not a set of behaviors and choices that a low sensation-seeker can easily relate to, but for the high sensation-seeker it is a delicious and absolutely essential feature of their lives, even their working lives.

One of the things I began wondering about along this journey was the working lives of high sensation-seekers. After all, I have a job that is arguably the lowest sensation career on the planet, or at least I thought I did. This led me to ask, "What kinds of jobs demand a high sensation-seeker and where are these people bound to thrive? What happens to the high sensation-seeking personality in a low sensation job like mine? And what would be an ideal place of work for each of the HSS subtypes?"

The answers baffled me. Jobs I would have thought of as a natural fit for high sensation-seekers actually turn out not to be. In addition, while sensation-seeking can be a helpful piece of the puzzle for extreme career paths like Jason's, in some cases it can be a hindrance, particularly if it is not tempered with a certain level of maturity, logical decision-making, and even empathy.

It was surprising that professions we classically think of as risky aren't necessarily the best fit for HSSs. Intuitively you wouldn't think this to be the case. Moreover, even in the rare cases where there is a fit between a risk-taking career and a high sensation-seeking personality, the correlation is far more nuanced than you'd think.

Risky Careers and High Sensation-Seeking

In 1977, two researchers out of the Illinois Institute for Technology, Robert Musolino and David Hershenson, wondered whether or not people who were employed in professions where risk is necessary also craved high sensation-seeking activities outside the workplace.[1] Maybe the risky jobs lead to relaxing after work activities like chilling out with a soothing television show. Or maybe those who love risk are drawn to risky professions. The implicit question was: Do risky occupations attract high sensation-seeking people and do HSSs tend to thrive in these jobs?

To find the answer, the scientists asked personnel training specialists to rank order ten career choices from highest to lowest risk. The researchers looked at work that involves placing themselves or others in jeopardy. The order from riskiest to least risky was as follows:

1. Test pilot
2. Air traffic controller
3. Policeman
4. Fireman
5. Psychologist

6. College professor
7. College student
8. Accountant
9. Civil service – clerical
10. Librarian

The investigators then compared the air traffic controllers with a group of college students and also people who work for the government. These air traffic controllers worked at Level IV facilities, the kinds of airports that handle over 300,000 flights per year. Being an air traffic controller means potentially putting many people in jeopardy in a rather stressful environment. What the researchers found was what you might suspect: The air traffic controllers outscored both groups across the board in sensation-seeking. So rather than kick back a break after a long, stressful day, it turns out most air traffic controllers are likely to prefer to seek out highly stimulating activities to wind down. This would lead one to believe that risk-taking professions and high sensation-seeking correlated. Case closed. Or is it?

The story actually isn't nearly as simple as that. Take pilots for example. Another study done in the 1970s showed that naval preflight students scored higher in thrill- and adventure-seeking but *lower* in experience-seeking, disinhibition, and boredom susceptibility than the college students.[2] On the face of it, this doesn't seem to make sense. You'd imagine that test pilots would score much higher in sensation-seeking than a college student. Even the authors suggested that the discrepancy between these results and the intuitive answer (that pilots would blow college kids out of the water on sensation-seeking) may have been due to "a social desirability response set" (research psychologist lingo for "the college kids wanted to sound cool" so they inflated their sensation-seeking scores). While this could be the case, when you think about it, it actually makes some sense that pilots would seek thrill and adventure but not necessarily score high in the other areas of the sensation-seeking assessment. Many test pilots are military personnel – the type of people who aren't likely to be disinhibited (that doesn't go over well with commanding officers). They also have to spend long hours in classes, in training, and in the air, performing highly technical movements under pressure over and over again, so if they didn't get bored easily or act impulsively this would be a major plus. And since they are likely to be following a relatively regimented lifestyle for years, sometimes decades on end, it stands to reason that they aren't exactly after new experiences. Instead, where they score high is precisely where you'd expect. These folks are after the thrill and adventure of flight. It would be a nice personality profile for this type of career.

As you move further down the list to police officers and firefighters, the relationship between risk-taking and high sensation-seeking only becomes subtler and even more complex.[3] Certainly, being a police officer or a firefighter means you are putting your life on the line and are therefore at risk, but it doesn't necessarily follow that high sensation-seekers are always the best fit for these jobs. Indeed, a study done in Poland showed that firefighters scored higher than everyday people only on the disinhibition scale (a trait one can easily see would be useful if you routinely were expected to run into burning buildings). In addition, a study that compared police officers to prison guards showed that police officers only scored higher in thrill- and adventure-seeking.[4] The prison guards (ostensibly a lower risk occupation) scored higher in each of the other four categories. From a common-sense standpoint, none of this seems to match up. After all, every night on TV we watch the latest show about first responders performing stunning, heroic deeds. I remember when *CSI* first went on the air and all of my students suddenly wanted to become investigators, imagining they would spend their days seeking out clues and cracking exotic cases. But the fact is, most police work is not like this at all. Patrols walk their beats or cruise in their cars often for hours or days at a time with little happening. Sure, there is bound to be the occasional chase or encounter, but having a low tolerance for boredom is certainly not one of the criteria for the job. In fact, many police officers spend much of their careers in the station filing paperwork or stuck on long, often boring stakeouts. For firefighters the story is pretty much the same. They might sit for days in the firehouse only to be called into action on a moment's notice. So being an experience and adventure-seeking type just doesn't match the profile.

I went to my local fire station to interview a few of the firefighters about their experiences and their sensation-seeking. There I met Firefighter Byrd. He remembers the conversation he had with his mother about choosing a career as a firefighter. He started to list off all the things he thought he wanted to do, hoping for her approval.

"Race car driver?"

"No," his mother responded.

"Police officer!"

"No."

"Pyrotechnic?"

"No!"

"How about firefighter?"

"That's all right."

So he went with firefighter. Oddly enough, being a firefighter is a bit like being a race car driver, police officer, and pyrotechnic all rolled in one. Byrd remembers his very first fire vividly. "It was two days after I got certified to go into fires. It was a residence where no one was home. It burned half of the residence down. Thankfully we got there just in the nick of time. Their furniture, their clothing, and all of the family pictures . . . we saved all of that," he said proudly. "It's a rush," he added, "it's almost surreal, because it's like you're scared because you know that this can go bad at any time. But you also know that this is the job that you have to get done."

It seems amazing to be able to rush into a chaotic situation and know exactly what to do. Training helps, I'm sure, but I wondered what stops him from freezing in those situations. "When that bell rings and it's time to go on a call, I'm not doing this for me," he explained. "Other people are depending on me. My crew is depending on me. Families are depending on me to save their life, homes, and their property. So I'm doing everything that I possibly can in order to protect that. And I feel like when you freeze, that's more about yourself. What *you* need. If you focus on a purpose greater than yourself, then you are able to move." And when you move you have to snap into action very quickly.

"It's awesome. You go from 0 to 60 because you literally go from being dead asleep to waking up and jump into these trucks where we don't know what we are going into. Sometimes you can see a big puff of smoke from a distance and that gets you even more excited because you don't know what you are getting yourself into but you know it is real. And you don't stop until after the fire is out."

Byrd scored 10 out of 10 on thrill- and adventure-seeking and his hobbies reflect that. Hiking, rock climbing, and he dreams of skydiving (not tandem because he wants to do it himself). On his days off he enjoys first-person shooter video games like *Call of Duty* and other stimulating kinds of fun. "I guess you can say I have an adventurous lifestyle."

"What's the biggest misunderstanding about firefighters?" I asked.

"That's easy. We don't get cats out of trees."

There are many other risky careers not included in Musolino and Hershenson's study. If being a test pilot, police officer, or firefighter doesn't require someone to be a high sensation-seeker across the board, what would? Maybe an emergency room nurse or doctor, an astronaut? Perhaps the military special forces?

The High Sensation-seeker in the Emergency Room

"It's exhilarating, I guess, for me. It's like torture, but, at the same time you just love it and want more of it. You know what I mean?"

This was Hannah's response when I asked her why she loved being an ER nurse so much.

"I think the people who stay working in ER, they definitely process it differently, like it energizes them maybe . . . Some people under stress they'll shut down. But, I think maybe people in the ER – the people who like it – they're at their best when they're under pressure, like athletes somehow."

It was beginning to sound like there was something to my hypothesis that emergency medical practitioners had to have a high sensation-seeking personality. After all, what environment could possibly be more varied and more challenging? A new health crisis presents itself at every turn, people's lives are at stake, and the decisions the ER technician makes in real time can mean life or death. Aside from BASE jumping into the Grand Canyon or hanging from a building by one hand it's hard to imagine an environment more filled with sensation.

Hannah certainly seemed to agree with me about this, and her life told that story. When I asked her if she engaged in high sensation-seeking activities outside of work she blithely replied, "I would say most of it is at work, I've done things in the past that were, I guess, thrill-seeking like snowboarding. I used to dive off the 10-meter platform. I did some skydiving, that kind of thing. I wouldn't say I'm an adrenaline junkie, but I've definitely done things like that in the past." Skydiving? Snowboarding? These are certainly high sensation-seeking activities in my view. I wasn't sure about her claim that she wasn't an adrenaline junkie either. When I pressed her for more, asking questions about what her friends did after work, I felt like the curtain was pulled back even further.

"There was a group in the ER I worked at that all did back-country snowboarding. I just loved the challenge of that. There were a lot of runners, people who did marathons and distance running. I think a lot of us do more extreme things in other areas of our lives also. I mean we don't all have the same personality, but after work, we would breakfast after a night shift and debrief it and just be like, 'Oh, my gosh, it was so crazy when this person came in and we had nowhere to put them and we had to pull that person to the hallway and we had to do a shoulder reduction in triage at the same time and . . . ' I think we like it and are always trying to top everyone else's story on what was the craziest thing that happened."

These do not sound like people who would like a desk job, and when I pointed this out to Hannah she heartily agreed. "Oh, no. Not at all. Like now, I work in a clinic because, over here in the UK, there weren't as many job opportunities, so I work in a clinic and I'm just bored because exciting things don't happen that often . . . I'm looking forward to going back to something more like critical care."

It was beginning to sound like I had found the perfect job for the HSS, but when I looked at the research, it didn't seem to corroborate my theory. A study done in 1987 showed that English medical students scored lower on the general sensation-seeking survey than arts and sciences and even agriculture students.[5] And while it has been shown that within the medical community emergency room physicians score higher than their counterparts who limit themselves to professional practice or teaching, it doesn't necessarily follow that ER physicians and nurses are inherently high sensation-seeking. This didn't seem to fit with what Hannah was telling me at all.

When I asked my colleague Dr. Philip Shayne, who runs the program at Emory University that trains emergency room clinicians, about the correlation between high sensation-seeking and success in the ER, his response was immediate and to the point: "That's part of it. Emergency medicine is a big, complicated field, and chaos is a normal for us. You need to be able to handle that, you need to be able to deal with a large amount of ambiguity and different information sources. We're sort of the interface between the chaos of the streets and the orderliness of the hospital. There are also moments of adrenaline that you need to be able to handle, and ideally you enjoy.

"On the other hand, I'm concerned about people whose primary focus is on that adrenaline rush. That needs to be a part of it, but that's a small part of what we do. In some ways, it's an easier part, a more scripted part. Satisfaction in this field really has to do with more long-term goals, about being part of a safety net, helping people out, being involved in system issues, being a generalist.

"You need a balance. You need to find people who've been tested in the environment and thrive in it. People who cannot multi-task or task switch, or think clearly in a crisis, or don't handle distractions well are not going to succeed in the ER. You have to be able to handle distractions very well. I need to know that people really have mature experience in the environment, and they've been assessed as doing well. If I read a professional statement that's all about the adrenaline rush, I'm worried about that. If they don't have a mature understanding, it's a career that'll lead to burnout if you don't really understand all aspects of it."

These comments were all given within the first minute of my conversation with Dr. Shayne. Clearly he had thought these issues through long before I called to interview him. I decided to dig in further and ask him what the right personality type was for an emergency room doctor.

While he was the first to admit that being a high sensation-seeker is a piece of the puzzle, there's more to it than that. He listed many criteria that were needed for people to excel in the ER. Some were what you might expect: the ability to make connections quickly, excellent communication skills, broad experience in the field. Some were less obvious. For example, Dr. Shayne described ER physicians as a "flawed type of doctor." He went on to explain that most doctors are able to have a great deal of precision around diagnosis, and this is an absolute necessity in most parts of the field. In the emergency room it's totally different. There's no time for that level of precision, not to mention the speed and chaos of the environment make it impossible. ER doctors have to paint things in broad strokes focusing on resuscitation, comfort care, and then risk assessment. "We're very different than other physicians. People who have something big and bad, get handed off. We're rarely definitive. We operate with a lot of uncertainty and a lack of precision. We're really meant to be the screen to chaos. You have to be comfortable being wrong. You have to have that sort of ego, or lack of ego to not mind being wrong. That's just different than a lot of doctors. Doctors don't like to be wrong." Oh, and you have to get used to getting yelled at a lot.

Being a high sensation-seeker can also cause problems in urgent situations. Let's say you work in an emergency room and you need to take a blood sample from a particularly squirmy patient. You are on a tight schedule and you have everything you need except one thing: the correct-sized gloves. You need medium gloves, but when you look in the supply closet there are only small ones. Rather than searching all around for the correct glove, you decide to gamble and wrangle your hands into the available gloves that are a bit of a tight squeeze. During the procedure the gloves break and you stick yourself with the needle. What's to blame? Your risky behavior or the gloves? Turns out that sensation-seeking not only affects what behaviors you are likely to engage in, but also how you justify the behaviors if they don't work out.[6]

How do high sensation-seeking medical professionals justify their unsafe actions in the workplace? They could either blame their own personality or they could blame the situation (sometimes called external self-justification), like hospital management or poor access to proper equipment.

Remember some high sensation-seekers don't see things around them as threatening, as average and low sensation-seekers do (it's that lack of cortisol). In addition, high sensation-seekers are less likely to think that their own actions will lead to negative outcomes. This is even more so for high-risk high sensation-seekers. We've seen that there's a difference between thrill-seekers and risk-takers. Risk is just what high sensation-seekers will tolerate for the thrilling experiences they want to have. Not all high sensation-seekers are risk-takers. But what happens when high sensation-seekers are drawn to risky behavior and the risk doesn't pay off?

To get to the bottom of this Martina Dwi Mustika and Chris Jackson of the University of South Wales in Sydney recruited 108 nurses and measured their risk-taking using a computer simulation called the Balloon Analogue Risk Task (or BART for short).[7] The BART measures risk-taking and involves pumping up a virtual balloon by pressing a button for cash. Every time you press the button you get a little cash and the balloon expands. You get to decide how many times to press the button, however the balloon could pop at any time and then you don't get any cash at all. You can stop pressing at any time and collect the money or keep going to maybe get more money. The test measures your risk tolerance or risk aversion. Risk was defined as the number of pumps. High scores indicated high risk-taking and pumping the balloon as much as possible. People who are risk averse cash out much sooner.

The researchers then asked the nurses to describe any potentially risky or unsafe actions they have done and to explain whether their potentially risky behavior was a result of lack of training and feedback, a lack of proper equipment, or simply a lack of time.

So what happens when risk-prone, high sensation-seekers' gambles don't pay off? Well they tend to blame the environment. When it comes to the workplace, high sensation-seekers who are tempted by risky decisions blame their unsafe behavior on lack of time or support. So, while sensation-seeking may help people tolerate the chaos of working as a first responder, when combined with a risk-taking personality, it may be more likely they'll take risks in procedures and then blame others when things don't work out like they hoped.

After my conversation with Dr. Shayne, I started to wonder whether or not I had been wrong, not necessarily about ER personnel but about my whole understanding of the relationship between high sensation-seeking and the workplace. I had thought that perhaps there was a certain type of job environment where the high sensation-seeking personality would excel. That didn't seem to be the case.

Looking back over my notes from my interview with Hannah, I realized she was hinting at some of the same ideas that came out of my conversation with Dr. Shayne. "You can't really tell just by someone's personality, I don't think, but you can tell how they act in a stressful situation. If they just jump right in and say, 'What can I do,' and . . . they might not know exactly what to do, but they're willing to do it and they don't freeze up, then you know they're going to be okay."

Certainly there is an HSS element in that kind of response. But there are many other things too. The HSS/career equation started to seem like it wasn't quite as linear as I wanted it to be.

One Small Step

When I was in Sudbury, Ontario, at Science North, preparing for a panel on which I was the psychologist discussing thrill-seeking,

I met Olathe, at the fossil table right outside of the meeting room. The fossil table is a demonstration area where you can teach kids to locate fossils scattered in a sea of flat grey pieces of limestone. The idea is that the kids learn to pick out the flat grey fossils from the flat grey stones. The activity might lead anyone with an average to high boredom susceptibility score to flip the entire table over in frustration. I was transfixed, sorting the stones, and stacking and balancing the discarded ones like a city of miniature Stonehenges. I noticed Olathe's laptop before I noticed her. It had two bright blue stickers on it – one for NASA and one for the CSA (the Canadian Space Agency).

"I know who you should interview for your book," she said even before introducing herself. I hear that a lot. I expected to hear about a neighborhood thrill-seeking skateboarder or ice athlete.

"Oh?" I asked.

"An astronaut trainee," she replied decidedly.

"Huh . . . " I thought. "Nice idea. But where in the world am I going to find an astronaut trainee?" Olathe smiled broadly.

I'm fascinated with people who are drawn to environments and careers that I would find absolutely intolerable. And as intolerable careers go space travel gets my vote. On TV and in the movies space travel is a joyous experience. It's smooth, sleek, and efficient. Passengers zoom around in antiseptic spacecrafts that look like class A office suites. It's even nice enough to bring your family and pets. For now, that's the stuff of science fiction. Currently accommodations for space travel are a little more clunky.

The early Apollo missions were jalopies compared to the Millennium Falcon and were relatively short treks. The fastest of these reached lunar orbit after only about two days. The Lunar Module spent another three days getting back to Earth. That's five days (if you don't count the time they spent on the surface of the moon) in a ship that's not much bigger than a Mini Cooper. Five days of roaring engines, no bathing, and having to do precise calculations.

Even in the more modern International Space Station (ISS), things are only a little better. The ISS is silent compared to older space vessels, but the background noise can be distracting, even befuddling. Air circulation fans, transfer pumps, and fluid coolant pumps produce a constantly shifting roll of noise, despite efforts to keep it as noiseless as possible. Over time, this noise can affect your hearing.[8] There's a relentless hum of about 69 decibels, which is about the loudness of the water when you are in the shower or when you are standing next to an average dishwasher (not the fancy silent ones).

If the sound of the space station doesn't distract you from your work, maybe the smell would. Retired NASA astronaut Scott Kelly, who spent an astounding 520 days, 10 hours in the ISS, likened the station smell to a jail, a "combination of antiseptic, garbage, and body odor. Mostly it's just exercise clothes people wear for a couple weeks without washing."[9] This is partly because when you are on the ISS you only change your socks and underwear every other day, and shirts and pants every ten days (if you're lucky). Plus if there is an odor there's really not anywhere for the smell to go. It's not like you can open a window.

Sounds awful. Lots of technical precise work to do, the hum of an eternal dishwasher and a lingering waft of 10-day old gym shorts. It makes traveling in coach seem luxurious. Who would crave, let alone sign up for this type of adventure? It's just the environment that high sensation-seekers like Olathe can handle.

Dr. Olathe MacIntyre is a scientist working with Science North's planetarium and space exhibits. Dr. MacIntyre is a biologist whose expertise is on growing plants in low-pressure environments – like the Matt Damon character in the book and movie *The Martian*. She's just the person you want to talk to if you want to grow a salad on Mars.

Those who know Olathe might not be too surprised she was interested in living in space. Inspired by her mother's science fiction collection featuring books by Arthur C. Clarke and Ursula K. LeGuin, Olathe has always dreamed of being an astronaut and she's had experiences to prepare her. She lived in Alaska for two years and spent a year on crab fishing boats, like the ones featured in Discovery's *Deadliest Catch*.[10] She works well in dangerous environments. I asked her if she ever felt like she was in danger.

"In danger?" She pondered for a moment almost as if she had to think about what the word meant. "Well, that's one of the things about me. I have a high threshold for not sensing being in danger. I worked up in Alaska for two years on fishing vessels as an on-board marine biologist also known as a 'fish cop.' It's a very intense environment. One of the first things you have to get used to is the motion. The first vessel I was on was actually a floating factory ship that had a crew of 150 men, mostly ex-convicts. On one boat you literally had to hold your plate and your cup when you were eating, otherwise they would slide off the table. It was so bad you'd learn to put your laundry in your bunk with you to kind of wedge yourself against the wall when you were trying to sleep. But there wasn't anything you could do about big swells, though. Sometimes you'd be sleeping and then you'd sort of wake up in a free-fall. There wasn't really anything you could do about that. You just kind of had to catch your sleep in between."

This is no cruise ship. These boats are smelly, noisy, rocking, and frenzied. It's not the laboratory environment that most biologists are used to.

"They process the fish on board, so it's actually a factory," Olathe explained. "They're actually fileting and freezing the fish on board, and it's tremendously noisy, so you're wearing ear plugs most of the time, ear plugs to sleep, ear plugs when you're working. You pretty much just need that protection the whole time because it's very very loud."

All of this while taking measurements, recording data, and sleeping whenever she could.

So when the Canadian Space Agency announced they were recruiting new astronauts, Olathe jumped at the chance. She wasn't the only one. Nearly 9,000 people started an application. Who signed up? Physicists, geologists, jet pilots. Some 3,771 submitted completed applications and were invited to do a government exam. Of those, 2,000 passed. After that Olathe was notified that she was in the top 160.[11] Then things got serious. She got a trainer, and started a rigorous fitness and nutrition routine. The possibility of being an astronaut made getting up at 6am, going to the gym, daily swims, and huge salads easier. She was training for her lifelong dream.

When she passed the medical exam, the number of candidates was down to 72 and she was selected for the next assessment: four days of intense tests, most of which she couldn't talk about.

"It was the most intense four days of my life. I can say that having been crab fishing in Alaska in the middle of winter. It started out with breakfast. Then we had to turn off our cell phones. There was no outside communication. You're not distracted. Your focus is complete. You're not thinking about work obligations or family obligations. Nothing. You're just doing this one thing.

"We were tested for flexibility, for strength, agility, being able to be a leader in a team, being able just as quickly to follow our spatial abilities, our spatial perception, our reaction speeds, our memory, learning, just being shown something and being able to do it again right away. One of the things we did that was really cool was we spent a morning at a pool. In the pool they had us do an obstacle course, like an aquatic obstacle course. Then there would be information on the bottom of the pool that we would have to retrieve. Then we had to remember and retain it. Then a couple of days later, do an exercise using the information of the pool."

Despite being stressful, Olathe's high sensation-seeking personality made the task easier. "In the pool I actually relaxed. I could tell I was doing well. I started actually having fun with my little team. We were laughing and joking. That made it easier to remember the information that I needed."

She got me thinking of something I hadn't really thought of before. If you ask an average or low sensation-seeking person if they would rather try to do something that they know they can do or something that they aren't sure they can do, many people would choose the thing they know they can accomplish. A lot of high sensation-seekers love the feeling of pushing themselves beyond what they think they can do. They thrive on the experience of pulling everything together in order to accomplish something. It creates an emotional and physical combination that's probably unlike any other.

Olathe agrees. "Yeah. It's a feeling of absolute presence. You're like 'Okay, I'm going to walk through that door. There's going to be some equipment ... I'm going to have to harness up to that climbing wall ... remember this number ... unscrew those bolts ... and get this moved over to there. That's all I need to do right now. I'm going to do it, and I'm going to work with these other people. We're going to figure it out.' It's a sense of empowerment action."

In the end, she wasn't selected. She got tripped up on a paper and pencil test. But her high sensation-seeking personality got her pretty far.

What did she learn about herself? "I learned that I get really intense. I knew I was intense. It's not like I had not been told I was intense, but it made me realize that I'm really intense. Even for people who are intense."

I wanted to continue to test my original theory however, so I decided I would seek out the most dangerous profession I could imagine and see how the high sensation-seeking personality expressed itself there. I wracked my brain and came up with a bunch of possibilities. There are a lot of dangerous occupations. However, I don't think anyone would disagree that being in the military Special Forces must be one of the riskiest jobs out there. Luckily, I had met someone who was both a high sensation-seeker and a retired military officer.

From Special Forces to Helping the Homeless

In his younger years, Corey was a classic high sensation-seeker. He rode motorcycles up steep winter trails that were frozen with ice, just for kicks. He dreamed of flying, bungee jumping, and skydiving. At some point he realized that there was one place where he could put all of his physical skill and high desire for intense sensation to work for him: the military. Being an intelligent, fit young man with little fear and a propensity for risk, Corey moved up the ranks quickly. He did airborne training. He was in air, sea, search, and rescue. Then he left that field to work in shock and trauma units like intensive care, critical care, and the emergency services. He had been engaged in some of the highest risk occupations in one of the highest risk fields in the world by his late twenties. Corey was clearly a high sensation-seeker, he knew it, and he put it to good use. When I asked him to tell me about a situation where he was able to use his HSS skills he responded without hesitation. "In search and rescue your primary goal is to go in behind enemy lines and recover pilots and/or anyone that they identify. When you drop-in your mind is really clear. You're wide awake and because you've trained so hard, you're in such good shape you can pick people up and carry them back if you need to. You have to do a medical assessment, take care of them in the field, prepare them, and bring them back."

Going behind enemy lines, rescuing fellow soldiers, doing medical assessments, and potentially carrying them out of danger on your back is difficult for most people to wrap their heads around. For Corey this was just another day at work. When he was finished with this part of his career he spent time in a military ICU unit where he specialized in pulmonary and cardiology issues. This is not like being in a typical ICU or emergency room. The problems are even more severe and immediate. Soldiers can come in with gaping wounds on the verge of bleeding to death and quick decisions have to be made. And because we are talking about war, procedures that aren't even allowed in civilian hospitals are routinely performed. In the military, you can do a lot of things you can't do in the civilian sector. "I used to assist at putting chest tubes in depending on where you were at. I put chest tubes in and then you sew the skin back around it. You almost get to the point where you look forward to it."

As a low sensation-seeker, I'd be easily overwhelmed in this line of work. I can't imagine having the calmness required to insert a tube into an injured person's chest and stitch them shut. I also can't imagine a situation in which being a high sensation-seeker could be more useful, more needed. Yet when I think of Corey's story I can see that being an HSS is only a part of the picture. Corey is a high sensation-seeker, but he is also strong, resilient, persistent, and intelligent, he has a strong ethical bent, and believes in helping others. Being a high sensation-seeker is just one aspect of his personality – albeit an important one – that allows him to thrive in such extreme conditions. He is a full human being as are all HSSs. Low sensation-seekers tend to think of high sensation-seekers as a kind of spectacle – partially because it's just so darned hard for us to understand them. Of course, they aren't. High sensation-seeking is just another trait and it can be positive or negative. It all depends on the person and the context.

Corey is well aware of this. Decades have passed and he made the transition back to civilian life successfully if not easily. He knows for a fact many of his high sensation-seeking buddies are struggling with this transition. When he returned to the states, Corey worked with vets coming home from military engagements. "A lot of my returning Operation Enduring Freedom men and women were like that. They were trying to reproduce that extremely rewarding combination. As a matter of fact, one of the guys that I worked with for a long time, he was a tanker, and he would say things like, 'They'd call us in when they couldn't get the bad guys out of a building. We'd roll up and take the building down.' There's nothing quite like having that power." It's hard to go from that to a desk job.

Interestingly, Corey has made this transition completely and now works in a relatively low sensation-seeking job – a mental health clinic where about 70 percent of the people he sees are homeless. He goes and seeks them out – it's like search and rescue only under city bridges and in subways instead of behind enemy lines – and brings them into the clinic where they receive food, shelter, mental health services, and help getting back on their feet. While these are clearly not high sensation-seeking activities, you can tell by talking to him that Corey gets a kind of thrill out of it. It may not be the same as inserting tubes into the chest of a severely injured person, but it does have deep meaning and purpose for him. Corey has taken his desire to help people, his integrity, and his fearlessness with him into his life in new ways, and has become something of a hero for it.

Risky Business

Does the personality and quest for thrill of sensation-seekers make them poor leaders? It's easy to think that high sensation-seekers may cause problems in business. Quite the contrary. Jayanthi Sunder, Shyam Sunder, and Jingjing Zhang from McGill University discovered that high sensation-seeking can have a positive impact on companies.[12]

Why is this the case? According to McGill University's Jingjing Zhang, sensation-seekers tend to embrace innovation and change, which could make them better CEOs. The researchers looked at the performance of 88 CEOs who were also pilots and 1,123 non-pilot CEOs in US firms over a period of a decade. What did they find? The companies with pilots as CEOs increased the number of patented products by 66.7, showing evidence of inspiring greater innovation. Their research suggests that businesses with a sensation-seeker CEO are likely to be more innovative. "Our research demonstrates that companies led by sensation-seekers, who display the same thrill-seeking tendencies as pilots, are able to generate more patents with greater market impact than their peers." Zhang explains, "This is because CEOs with this particular personality typically improve innovation effectiveness and pursue more diverse and original projects."

"Managers with an inclination for creativity in corporate settings are far more successful when innovating. An openness to new ideas, and a willingness to pursue new methods of working overrides their desire to maintain structured and repetitive situations. They are also likely to be more innovative consumers, unafraid to try new products and always aware of alternatives," she explains.[13]

Remember Munir from Chapter 3? He's my former student who tricked his then girlfriend into eating goat brain. He said his high sensation-seeking personality influences his work and career decisions.

"I think it comes through in my work, as well. Like with corporate America, I'll get bored very easily because there's no more thrill-seeking. It's interesting when you were talking earlier about people who seek chaos and thrive. I think I'm very much like that. I think I need to have like 10 or 20 things going on at the same time to be able to almost be successful. Like if it's just one thing, I would get bored. If I've got multiple things, multiple problems, I thrive in situations like that. I think all throughout college I was like that, as well. If I've got to write a paper and I've got three weeks to do it, it's too boring for me to be able to have the motivation to do it. But if, I need to do this now, it's like okay, let's do it. I think it's interesting because I enjoy new experiences. I scored high on the thrill-seeking, but not, super high. It's just that I always want new experiences and I'm not afraid of risk. I've left four corporate jobs within four years of graduating because I literally get bored to the point where I think, you know what? I could do better than this, and I move on. So it is very interesting when you said you got the self-trust. You got the fearlessness. I leave jobs with literally just the self-trust of I'm going to find something better because I know I'm worth it. In that sense, I've got the thrill-seeking down."

~

Talking to these high sensation-seekers about their work life and looking at the research led me to a new and unexpected understanding about the interface between the HSS and the workplace. In some cases, it is probably true that high sensation-seeking is an important pre-requisite for job performance and satisfaction. Jason (the underwater Indiana Jones) is a great example of this. I can tell you from personal experience that a low sensation-seeker would not only loathe a job that required them to tie heavy weights to their bodies and sink to the depths of black water during hurricane season, but they also flat out wouldn't last long in the position. But how many jobs can you think of where this kind of extreme behavior is required? Not many.

It is more often true that the *types* of sensations an HSS seeks are connected to job satisfaction. A sous chef in the latest nose to tail restaurant almost certainly has some experience-seeking tendencies, you'd have to be adventurous about what you eat. And the local circus just wouldn't be nearly as fun if the clowns weren't a bit disinhibited.

The vast majority of us simply don't work in careers like these, and that includes many HSSs. After all, even jobs we perceive to be filled with stimulation often turn out not to be in real life. Sure, being a firefighter is an intense career, but not every call is a four alarm fire. There are small kitchen fires to douse, people stuck in elevators to rescue, and of course, reports to write, and maybe even a cat or two to rescue.

Most HSSs know this, and they don't expect their job to fulfill the part of themselves that desires intense experience. Instead, they seek that out recreationally. Like Indiana Jones, they masquerade as normal folks during the work week, and then use their weekends to go spelunking, BASE jumping, do Tough Mudder competitions or eat exotic foods most of us would never consider trying.

However, there is an important lesson to be learned here. High sensation-seekers who become firefighters, police officers, test pilots, or get into careers we often think of as risky in the hopes that their days will be filled with intense experiences could find themselves bored, stifled, disappointed, and may underperform because they are not being challenged.

At the end of the day, there is no perfect HSS job, but there are many scenarios where being a high sensation-seeker can be an asset. If you're an HSS, it may be more likely you will thrive in a chaotic environment such as the emergency room, but it doesn't necessarily predict your success there. There are many other factors against which the high sensation-seeking personality must be balanced if he or she is going to flourish in any given career. Like many things, being an HSS can be a blessing or a curse. It all depends on the broader context of the work, the environment, and the character of a given person. In some cases, it can be like a super power. In others it can take a dark turn and lead people down a path no one wants to go. It's time for us to walk this dark path and see what happens to the high sensation-seeker when the need to feel the buzz goes too far.

7 THE DARK SIDE OF HIGH SENSATION-SEEKING

Daedalus, a brilliant Athenian architect, was commissioned by King Minos of Crete to design an elaborate and complicated labyrinth. The resulting maze was so tricky and confusing that once you were inside, it was nearly impossible to navigate without assistance from the gods themselves. Then, to punish Daedalus for providing a secret clue to escape the labyrinth, the King had both him and his son Icarus imprisoned. But it's hard to keep a good man down. Daedalus and his son escaped and hid on the shore near the rocky cliffs of Crete overlooking the sea. While hiding, Daedalus looked longingly at the seagulls that circled the beaches. Too bad he and Icarus couldn't just fly off like the birds – unless... Daedalus decided to build two pairs of massive seagull wings so he and his son could escape.

While Daedalus designed the wings, Icarus was to gather supplies. Little by little, Icarus collected thousands of seagull feathers, branches, twine, and pounds of beeswax for the project. Daedalus fashioned the wings, bound them together with the thread and molded them with the beeswax.

Before their escape, Daedalus went over the plan. "Now this is important," he warned. "Follow my EXACT path. If you fly too low the sea spray will weigh you down. If you fly too high the sun will liquefy the wax and your wings will fall apart. In either case, you'll crash into the sea."

They strapped on their wings. Daedalus charged toward the sea and extended his arms in the enormous cape of feathers. Icarus followed. With an unfurling whoosh, Icarus was airborne. He was actually flying! His body was weightless. He was free. He felt the fresh salty sea air on his face, the rush of the wind in his ears. Higher and higher, faster and faster he flew. Free of the ground and free of the prison. All he could see was the pale blue sky and the deep blue of the ocean and in the distance a small speck below him. It was his father. Then he noticed a tiny feather drift past him and float away; then another. One by one the feathers scattered, like dandelion seeds in the wind. The blazing sun softened the beeswax and the wings were falling apart. Before he knew it, his arms were bare, and Icarus crashed into the sea.

The story of Icarus is a lesson in moderation: Don't fly too high and don't fly too low, just keep to the middle. But can you really blame Icarus? After all, he just escaped from an island prison and was flying for the first time. Moderation isn't always the primary strategy for sensation-seekers either. Many high sensation-seekers live their lives in extremes and there are times when abandoning moderation leads to wonderful experiences. But at other times, you end up waxed and feathered, drowning in the sea.

Despite the extreme stories and cases, for most people, sensation-seeking isn't a problem at all. The vast majority of sensation-seekers will tell you that sensation-seeking is a valuable part of their life. Their thrill-seeking experiences make the world seem bigger, closer, and more amazing. Being a sensation-seeker can even be an important part of their work (as illustrated in the last chapter).

However, some people with sensation-seeking personalities struggle with the dark side of risk-taking. Sometimes sensation-seeking can wreak havoc in the form of making the wrong decisions for the sake of a thrill, or even falling into addictions and aggression. In some cases, sensation-seekers don't even know they are getting into trouble until it's too late. Interestingly, some of these problems can be predicted early in a person's life. Throughout my research I came across examples of how sensation-seeking can lead to trouble which led to a better understanding of how tempting Icarus with a marshmallow might have predicted his plunge into the sea.

Worth the Risk?

Wes, a 35-year-old marketing executive, pursues a lot of hobbies and interests: mountain climbing, skydiving, mud runs, hiking. He learned early on in his high sensation experiences that "adventure-seeking . . . takes you right to the edge."

"I think it started when I was 11 years old," Wes explained. "My family and I were taking a trip to California. The plane lost cabin pressure and nose-dived 30,000 feet. I fell unconscious and perforated both of my eardrums. It was one of the most painful things I've ever experienced. I remember looking out the window seeing death happening as we nose-dived. Ever since then, when I'm faced with some kind of adversity, I often try to challenge myself to do something out of the ordinary."

As I've mentioned in previous chapters, early life experiences may set the stage for sensation-seeking to develop. Nose-diving 30,000 feet as a child would clearly create a lot of sensation. It may be that Wes had a higher tolerance in his adult life for situations most of us would experience as stressful, since he faced death at such a young age. Whatever the case, Wes clearly embraced being a sensation-seeker.

In some cases, this was a good thing. Having been in a nearly fatal accident so early in life led Wes to pursue activities like mountain climbing and skydiving which, in turn, helped him conquer his fears. But even though these experiences usually helped him, he discovered that sometimes they went too far. One such experience illustrated how his sensation-seeking was spiraling out of control.

"When I was in high school, I was big into hiking and mountain climbing," he explained. "I decided I wanted to climb a volcano – Cotopaxi. It's at an elevation of around 19,000 feet. I wanted to see what sunrise at the top of an ice-capped volcano looking down on the clouds was like. I thought, 'If can do this, I can do anything.'

"So, I flew to Quito, Ecuador. From the airport I drove to a little store to pick up a tent and some gear, and then went straight up to Cotopaxi National Park and started hiking for hours. Going to the mountain immediately after landing is probably one of the stupidest mistakes I've ever made. I got to the first station where a lot of climbers stop to eat and chill, and I noticed I had a little headache. I figured it was no big deal. But by the time I got to the second station, around 15,500 feet, I knew I was in trouble. I could hardly breathe and thought, 'Why am I huffing and puffing? I'm in great shape. I run all the time. What's going on?' Later in the evening, I began hyperventilating. Everybody went to bed at 8pm to start the trek up the volcano at midnight and I kept telling myself, 'I can do this. I can do this,' but my heart was racing.

"The next thing I remember I woke up vomiting. I had to swallow my own vomit, because I was on the third bunk and I didn't want to puke on everyone below me. I had an unbearable headache. It was so bad, my eyes were tearing. Right around midnight, we all went outside to start the hike. Someone mentioned something about my face being blue, but I ignored it. I got all my gear on, and was going to start the hike anyway.

"I remember looking up at the stars. I've never seen a sight in my life like it. So many stars. It was the most beautiful thing I've ever seen. At that moment I had to decide, do I continue? Because I knew I would die if I did. I had a little medical background, and deep down I knew I was in really rough shape. My face was bloating, I was turning blue, and I was vomiting. Even so, I was debating internally about trying to make the climb. I had spent a lot of money on the trip, and didn't want it to go to waste. I also didn't want to disappoint my dad, who has always been really proud of my travel. On reflection, this was stupid, but at the time it seemed important to me.

"Everyone started heading up the mountain, and that's when I ran into a woman who saw me and said, 'You're cyanotic. You're turning blue. When did you get to Ecuador?' 'Today,' I replied. Her mouth dropped. 'You need to get off this mountain! Now!'"

Wes was experiencing "mountain sickness," also called hypobaropathy, or the altitude bends – the result of low oxygen at high elevations. Mountain sickness can occur at altitudes as low as 8,000 feet, and it gets worse with exertion and with rapid altitude changes, so it's no wonder Wes was feeling sick at nearly 15,000 feet above sea level. His body was desperately scrambling for the scarce oxygen in his blood. Mountain sickness can progress quickly to high-altitude pulmonary edema, which can be fatal within 24 hours. This frightening condition can involve difficulty breathing, weakness, and cyanosis (turning blue) because of low oxygen levels. Luckily, mountain sickness is pretty rare, occurring in less than 1 percent of people.

Wes abandoned his quest to see the sunrise on Cotopaxi and left the camp. "I convulsed in a hotel in Quito, but finally the next day I was okay. It made me realize that adventure-seeking sometimes takes you right to the edge."

Think back to Zuckerman's definition of sensation-seeking from Chapter 1: "sensation-seeking is a personality trait defined by the search for experiences and feelings that are varied, novel, complex and intense and by the readiness to take physical, social, legal, and financial risks for the sake of such experiences."[1] For Wes the quest to see the sunrise on Cotopaxi outweighed the risks. Luckily he learned from that experience.

"That moment shaped who I am and everything I do now," he explained to me. "I always plan. If I go to another country, I ask questions. What vaccines do I need? What's the altitude? A quick example: Two years ago I was supposed to go to Machu Picchu with a friend of mine, Ricky. I planned everything out, because he's from Ecuador, but he's a flight attendant and lives in Dubai. At the last minute, he couldn't go. I'm all packed, but I said, 'I'm not going.' It wasn't that I didn't want to do it myself per se, it was that because I did my research. Cusco and Agua Caliente are at such high altitudes that in hotel rooms, they have oxygen tanks. When people do these trips, they usually go in pairs, especially at that altitude because you have somebody who can say, 'Hey, you're not looking very well.' It was like this was taking me full circle back to the camp at Cotopaxi, and this time I knew I had to make a different decision."

Risk on the Road

We have all seen the signs on the side of the road: "Speed Limit 60." Some people take this seriously, as if the sign controls their foot and won't let the car travel even a tiny bit above the limit. Others see it as a mild suggestion: "If you are wondering what speed to travel, please consider 60." Still others act as if the sign says, "Minimum Speed 60." Nearly two-thirds of us travel at speeds well beyond the posted limits.

How do we choose our speed? Each of us has a zone in which we feel most comfortable when driving. Too fast makes us feel that we're driving dangerously; too slow frustrates us. We've all seen "that guy" who weaves between cars as fast as possible, maybe even flashing high beams to warn those ahead that he has no intention of slowing down. What makes that person feel that their behavior is safe?

Maybe they are a high sensation-seeker.

As you might imagine, sensation-seeking is linked to risky and aggressive driving behavior. When Marvin Zuckerman and

Michael Neeb asked people to estimate how fast they usually drive if the posted speed limit is 55, those with a total sensation-seeking score in the medium range of 20 said that they tend to respect the posted speed limit. However, crank that sensation-seeking score up to the mid-30s and suddenly you have a different story. When the sensation-seeking score was 34, people reported zooming at 75 miles per hour.[2] What is it that drives sensation-seekers' pedal to the metal? High sensation-seekers find thrill and adventure in speeding, following the car in front of them too closely, and other risky driving behaviors.

One self-identified sensation-seeker, Sue, explained that driving fast was fun and made her feel more alive. Sue said that her friends often made fun of her car. It has so many dents and scratches that you might expect a crash test dummy to emerge from the driver's side. Sue had been in several fender benders but assured me that she was a skillful driver – especially when she was driving fast.

"All my friends say that I drive like a race car driver," she explained. "I feel like I actually drive better when I'm not necessarily following all the rules. One thing I realized is that I love curves. I really like to start in one lane and then to change across three lanes of traffic if there's an opening. What I like about it is that I feel extremely focused. Even when there's a close call . . . I feel like it's an accomplishment.

"I like speeding. I really like weaving in traffic. I like how smooth it feels and how focused I am. I like getting ahead of the cars, it's like a game. I take turns very quickly. I pretty routinely run yellows. I think it freaks people out."

I bet. It would freak me out, and that's because I have a very different risk tolerance than Sue does. All this time I figured when I saw someone cut across three lanes they must have suddenly realized that they needed to be three lanes away. It never occurred to me that for some people, driving is a game.

Even the best drivers might admit that driving is never completely safe. Not much is. We face risks all the time. As everyday risks go, driving is objectively a big one. We routinely get into a large metal object on wheels, operated by our hands and feet, propelled by an over-powered motor that can easily go twice as fast as the speed limit, on narrow patches of road with only three feet of space on each side. We do so amid scores of other people doing the same thing, many of them pressed for time and distracted by texting, putting on makeup, scolding kids, reading, or swatting bees. It's a small wonder anyone gets from one place to the next without running off the road or into something along the way. Unfortunately, in any given year, thousands of Americans are seriously injured in automobile accidents and over 40,000 lose their lives.[3]

But sensation-seeking influences how we approach those risks. Most of us have an amount of acceptable risk we feel comfortable with – an idea called "risk homeostasis theory."[4] When the perceived risk rises above a certain level, we suddenly become more cautious. When it falls below this level, we feel more comfortable and will take more risks. This might explain why we tend to be more cautious when we are walking or driving in unfamiliar areas or at night when we can't see very far.

Another model, called the "zero risk model," suggests that most of the time we don't find our activities to be risky.[5] When we do, we begin to be more careful. These two different models essentially say the same thing: When our perception of risk gets too high, we change our behaviors accordingly. Whatever theory you favor, one thing is certain: The perception of risk varies from person to person. HSSs may very well be greater risk-takers, because they don't perceive their behaviors to be all that risky in the first place.

High sensation-seekers may take these driving risks to avoid boredom. Sue with the dented car says that she only crashes when her husband makes her drive "normally." "I get sleepy when in traffic, when I'm driving slow and following the rules. When I'm driving residentially, that's when it's kind of boring and I zone out and that's when I crash."

She's not the only one. Monotonous conditions are particularly hard for high sensation-seekers. In a study of highly monotonous road conditions, sensation-seekers tended to either become extremely restless or fall asleep.[6] It might be important for high sensation-seekers to find safe ways to keep entertained on long road trips. Anything that's safe that might provide a bit of entertainment like singing, chatting with passengers, or even car games might be a better alternative to zipping and darting from lane to lane.

Some people drive fast because they like the way it feels, others do it to avoid feeling bored, but there's another perceptual reason people speed, despite the inherent risks of doing so: it's that they falsely believe that travel time decreases as driving speed increases – it does obviously, but not as much as you might think. About a third of drivers report speeding because of time pressure, like being late for an appointment.[7] However, we tend to overestimate how much time we will save by driving faster. On a trip I took from Macon, Georgia, to Atlanta, my GPS display predicted that the drive would take 90 minutes. I figured that if I drove just a bit faster than the 65 miles per hour speed limit, I would get to Atlanta sooner. As I pressed the accelerator down, my speed crept up to 70, at times maybe 75. Okay, 80. Strangely, the GPS display didn't refresh and show a new estimated time. I thought maybe the GPS hadn't calculated my new speed yet, and that maybe it just took a while to refresh. Nope. I arrived in Atlanta about 90 minutes after I left Macon. What gives? The estimated time of arrival actually doesn't change that much, despite the fact that you are driving faster. This miscalculation is called "the time-saving bias."[8]

Let's break it down. Let's say you live 10 miles from work and you take a local road to get there. If the speed limit on that road is 20 miles per hour, it will take you about 30 minutes to get to work. If you increase your speed to 30 miles per hour, you can get there in only 20 minutes, a whopping 10 minutes faster. But if your job is 10 miles down a highway where the speed limit is 50, you have a different situation. The same 10 miles per hour increase actually results in less time saved if your initial speed is higher. Zooming 60 miles per hour gets you there only two minutes faster. It's not exactly worth getting a speeding ticket to shave two minutes off your journey.

High sensation-seeking individuals tend to be particularly susceptible to this time-saving bias. This bias, together with their thrill-seeking and boredom susceptibility, leads high sensation-seekers generally to choose higher speeds than their low and average sensation-seeker counterparts.[9] Some research suggests that fast driving can produce thrill in those who score high in thrill- and adventure-seeking but also can be an illusion of control that results in feelings of superiority for some drivers. Ask around, some people are really proud of their driving prowess, no matter how dented their cars might be.

I learned in Drivers' Ed in high school that two car lengths from the car in front of you was a safe following distance. You needed to keep at least that distance because you never knew what could emerge ahead. The car you're following could stop suddenly, its driver could toss things out the window, a rabbit could dart out in front of your car – all sorts of unimagined chaos. The best way to avoid it is to keep a safe distance.

But most people have their own internal set of rules about following the car ahead of them. Have you ever wondered why some people are perfectly content to leave a football field's worth of space between them and the car ahead of them, while others seem to be satisfied only if they are nearly in the back seat of that car? Sensation-seeking might offer a clue.

As a low sensation-seeker, close following causes me to scramble for the invisible passenger emergency floor brake or to jokingly announce, "We have obtained ramming speed!" to my friend who approaches cars from behind with aggressive enthusiasm. If you ever wondered how your high sensation-seeking friends tolerated that much stress, the answer is they don't even feel it. Researchers observed the driving behavior of subjects as they rated their risk feelings at various points in time as they followed another car. The researchers also kept track of different stress indicators such as heart rate. High sensation-seekers followed the lead car at closer distances but didn't perceive their tailgating as more risky. The low sensation-seekers showed a greater increase in perceived risk and a threefold increase in heart rate.[10]

The novelty and intensity of stimuli produced by speed and reckless driving release dopamine in reward areas of the brain in high sensation-seeking individuals, as we discussed earlier. The thing is, almost everyone drives with the same amount of perceived stress. We choose the distance where we feel comfortable, just as the risk homeostasis theory predicts.

Before you feel too comfortable about the high sensation-seeker's driving powers, there are a slew of other driving facts to consider regarding high sensation-seeking drivers. Not only do those with high sensation-seeking personalities drive faster and closer, but they also have more accidents and convictions for driving offenses. Sensation-seeking is correlated with fast and careless driving,[11] increased multitasking and distracted driving,[12] speeding and other traffic violation convictions, and accidents.[13] HSSs tend to wear their seat belts less often,[14] race other drivers, pass in no passing zones, and try to beat trains at train crossings.[15] In fact, over half of high sensation-seekers have one or more accidents, while about a quarter of low sensation-seekers have never had one (raises hand).[16]

If It's Worth Doing: Addictions

Sensation-Seeking and Substance Use

Maybe you've heard of the term "addictive personality." People with addictive personalities tend to get locked into a pattern where they overdo nearly everything, throwing moderation to the wind. For them, anything worth doing is certainly worth overdoing. While many people like playing video games for a while, people with addictive personalities will play for ten hours in a row. When they discover a new TV show, they'll binge-watch five seasons in one weekend. They easily fall into things that feel good, like gambling, sex, and sometimes drugs.[17]

Some people with addictive personalities don't limit themselves to overdoing just one thing – for them, an addiction to one thing is linked to an addiction with another. Psychologists refer to this as "cross addiction." People addicted to alcohol, for example, have cross addictions to cigarette smoking, overeating, and gambling more often than people who never abuse alcohol. Sensation-seekers may be more likely to be drawn into addictive substances and even into addictive behaviors.

Devin, a 32-year-old man living in south Florida, knows something about sensation-seeking and addiction. He's a resident manager in an addiction treatment facility, a sensation-seeker, and a recovering addict. He's known he was a thrill-seeker for decades.

"I could probably go back to being like six or seven years old and reading. I can't think of the book exactly, but it was one of those 'choose your own adventure' style books. The kind where 'if you want to go here, go to page 84; if you want this to happen, go to page 78.' I remember actually having my heart race as I'm reading those books in anticipation of what was going to happen." Devin's thrill-seeking went beyond reading.

"Even when I was a kid playing sports, skateboarding, riding bikes, I always tried to do something more extreme than the person next to me. People like me just don't find the same satisfaction in normal things that other people do. It has to be more."

As Devin got older, he sought different sensations including loud music, roller coasters, and driving as fast as he could. He began using marijuana around age 12. Later, he added alcohol, then cocaine and opiates like heroin. "It was a natural progression. For me, it was almost like I knew what the end result was going to be, but I didn't care about the consequences. It just didn't matter in that moment."

It's hard to imagine that anyone gets involved with substances intending to get hooked, much less to have the substances slowly destroy their lives. Lots of bad ideas start out with elements of good ideas. Sometimes people smoke, drink, or try drugs to solve problems like boredom or psychological or physical pain. These kinds of feelings can be overwhelming and unbearable. Doing something that eliminates unwanted feeling can be tempting, but it often comes with a price. Long-term addiction can create problems with friends, family, work, and physical or financial health. Some people experience a downward spiral that's difficult to escape or reverse.

Explaining what might motivate a person to use drugs is complicated, involving complex social, psychological, and environmental factors that many experts and volumes of books have addressed. A full review of this research is beyond the scope of this chapter, but it is relevant to give a glimpse into the relationship between sensation-seeking and substance use and abuse. Sensation-seekers, as you know by now, are more likely to be drawn to unusual experiences and much less likely to be fearful of risks. In their examination of substance use and sensation-seeking, Bernard Segal and his colleagues found that sensation-seeking was associated with substance use.[18] It would seem that, in some cases, HSSs' quest for the buzz goes beyond BASE jumping and cliff diving. On average, the higher someone's sensation-seeking score, the more substances they're likely to have tried.[19] Sensation-seeking scores are lowest for those who've never tried any substance, a bit higher for those who've used only alcohol, and highest for those who have tried illegal or multiple substances. What's more, Andrea Kopstein and her team surveyed some 1,100 eighth graders and over 1,200 eleventh graders and discovered that those who score high in disinhibition are three times more likely to smoke and six times more likely to have tried marijuana than those who score low in disinhibition.[20]

Though being a sensation-seeker is linked to trying substances, it doesn't appear to predict the type of substance a person might use. Sensation-seeking seems to be more related to the number of substances a person has tried rather than what they tried. As Zuckerman points out "sensation-seekers like to get high and they like to get low."[21] It's not that they set out to get addicted – obviously. No one does. Rather, it's that their need to seek out unusual experiences drives them to try dangerous things. And sometimes it just doesn't work out so well. Zuckerman found that curiosity, pleasure, and avoidance motivate sensation-seekers to use addictive substances, whether tobacco, alcohol, or illegal drugs such as cocaine or heroin.

CURIOSITY

Nearly everyone, at some point in their life, has wanted to feel something different than what they were feeling at the moment. Maybe they've wanted to chase away the morning fog, or escape stress, or feel excitement or joy instead of being down in the dumps. Or maybe they've wanted to enhance the way they're already feeling. Imagine how these feelings might be intensified for the sensation-seeker. Imagine the profound curiosity they may feel about how LSD would make them feel or how cocaine might break up the boredom. This curiosity phase of substance use involves wanting to instantly summon that desired feeling. Even without having tried any of these substances before, many of us have seen or heard about their effects on others – for example, the relief on someone's face when they take a drag from a cigarette, or TV and movie depictions of the excitement of a night of club drugs. You may have found yourself thinking, "I wonder how THAT feels?" That's the curiosity phase. Still, while a lot of people are curious, they won't try these drugs. After all, cocaine, heroin, and methamphetamines don't have the best reputations as life-improvement elixirs.

For some sensation-seekers, though, curiosity wins out over the potential risks. Devin was curious about the effects of drugs, and the potential risks didn't persuade him to stay away from them. "I knew how real the consequences were, and I've seen other people before me go down the same path, but I rationalized it. I told myself I wouldn't get that bad or I wouldn't end up like that." As Zuckerman found, sensation-seekers are more apt to give in to their curiosity about drugs.

Social factors are also related to substance-use curiosity.[22] Our attitudes and beliefs about substances are influenced, in part, by who we hang out with. I never saw either of my parents go near an alcoholic beverage. My parents seemed convinced that, in an instant, you could get hooked on nearly anything, including Advil. I remember my dad requiring me to explain to my third grade teacher that my orange Tic Tac capsule-shaped candies were not, in fact, drugs, lest she suspect I was popping tangerine colored pills between social studies and spelling. I was raised to be terrified of drugs, and so my well-learned terror overrode any curiosity that may have bubbled up. This is quite different from my friend in elementary school, who seemed weaned on beer and Jägermeister before he was allowed to cross the street without adult supervision. Where my family and I saw substance use as highly risky, my friend's more relaxed attitude was probably influenced by having sensation-seekers among his family and other friends. To these sensation-seekers, everything, including substance use, seemed less risky than it did to me. This doesn't mean that my friend grew up in a culture that encouraged drug use. Instead, the barrier between curiosity and use for him may have simply been a bit thinner.

FROM CURIOSITY TO PLEASURE

Getting past the curiosity leads to trying the drug and the next motivational phase: pleasure. The effects of many psychoactive drugs on the neurotransmitter dopamine make using these drugs feel good. Dopamine, as you may remember from Chapter 2, is a neurotransmitter or chemical messenger involved in experiences of pleasure. It's dopamine that's involved in your reward center, reinforcing natural things such as laughter, sex, and grooming behavior like combing your hair and even picking your nose. When dopamine is released, you feel good.

While the natural release of dopamine induces good feelings, substances such as methamphetamine interact with dopamine to produce pleasurable feelings in a very different way. Because methamphetamine resembles dopamine in the body, it is easily absorbed by neurons. Methamphetamine molecules slip into the neurons and displace the naturally occurring dopamine. The dopamine receptors become overstimulated when the displaced dopamine becomes trapped in the synapse, causing the neuron to bind to the dopamine again and again. This overstimulation creates a sense of pleasure.

Pleasure is reinforcing to all of us. However, it may even be more impactful on people with high sensation-seeking personalities. As you know from previous chapters, HSSs tend to release more dopamine in high sensation experiences. So it may be that drug use is even more pleasurable for them than it is for average or low sensation-seekers.

FROM PLEASURE TO AVOIDANCE

People try drugs like methamphetamine out of curiosity, to see if it makes them feel the way they've heard or hope to feel. But what starts the behavior is different from what keeps it going. They have to use more and more of the drug to bring about the pleasurable feelings it initially created. That's because the pleasure of dopamine overstimulation comes with a price. Blare your speakers too loud for too long, and they'll probably burn out. Dopamine receptors can burn out, too. When they do burn out, people need more dopamine to be released in order to get an average amount of pleasure. This change, known as tolerance, means they need more and more of the drug to get the same effect. It can also mean they feel miserable when they aren't under the influence of the drug.

In this "avoidance" phase, people use the drug to prevent bad feelings of withdrawal. For Devin, things got much worse as tolerance kicked in. "I didn't necessarily want to stop using. I had everything going for me that an addict/alcoholic could want. I made plenty of money doing bartending at concert venues, I saw live music every night, and I made cash. I sold my drugs of choice so I never ran out. But the drugs and alcohol just stopped working, and I got very depressed. I got to the point where I couldn't see myself continuing the way I was, and I couldn't see myself stopping. My wife and I got separated, I had a few good friends pass away or take their own lives, and I just got really bad really quick ... I tried to overdose. I actually tried to kill myself."

After his suicide attempt, Devin woke up in a hospital. "At the time, I thought that waking up in the hospital was the worst thing that could ever happen. I wondered if my pain was ever going to end. I just wanted everything to stop. I ended up in a treatment center for 55 days. Then I remember one day I noticed palm trees for the first time. Before I was just numb and dull to everything. Because of the drugs, I hadn't noticed them. That's when I was really grateful that things worked out the way they did."

Things are much better for Devin these days. As a resident manager in an addiction treatment facility, he says that he can recognize unhealthy sensation-seeking in his residents. Knowing how they experience the world allows him to help them. Sensation-seeking, he says, has been both a help and a hindrance to his own recovery.

"I know I can act impulsively. It's more like I don't really worry about the consequences – 'we'll sort them out later' – and that's not always the best way of thinking. But [sensation-seeking means] being able to . . . create that happiness for myself, or make myself feel good through whatever. It may be a benefit, because I see a way to enjoy my life without having to get high. And I still find satisfaction in doing thrill-seeking things."

Behavioral Addictions

People can be addicted not just to drugs, but also to behaviors. Researchers have found that sensation-seekers can struggle with behaviors they do too much, creating problems with friends, family, and work.

Gambling is one sensation-seeking behavior that can become addictive. For most people, gambling is just a fun diversion. They gamble responsibly, with only slightly smaller bank accounts to show for their efforts. But there's a difference between a retiree at the penny slots and someone who takes big risks. People with a gambling disorder seek intense stimulation, excitement, and change, and they love risks.[23] Even in laboratory studies of lottery strategies, sensation-seeking was associated with riskier gambling practices.[24] Nadia Kuley and Durand Jacobs studied problem gamblers outside of treatment and found that they scored higher than social gamblers on total sensation-seeking.[25] In fact, the higher your scores on disinhibition and experience-seeking, the larger percent of your income you are likely to spend on gambling.

Pathological gamblers develop a preoccupation with gambling that can devastate their families' financial wellbeing. For these individuals, gambling becomes a behavioral addiction that is linked with significant problems with family and work. They grow concerned with the act of gambling to such a degree that they put their personal relationships, occupational status, and financial stability at risk.[26] They continue to gamble despite increasing family and financial distress, have difficulty controlling the urge to gamble, and are often unable to reduce the amount of time and money they spend gambling. Eventually they don't get the same thrill from gambling so they have to increase their bets or the frequency of gambling to get the same effect. They often become irritable if they cut down, gamble to lift their mood, lie to hide how much money they spend, and rely on others to bail them out. They also frequently gamble with progressively larger amounts of money in order to meet their ever-increasing need for excitement and pleasure.

Stephen McDaniel and Marvin Zuckerman noted that high sensation-seekers had an increased interest in gambling, had larger bet sizes, were more likely to try to "make up" losses by raising bet sizes, gambled longer despite losing more money, and that sensation-seeking score varies directly with the number of gambling activities, bet sizes, and money spent on gambling.[27] Pathological gamblers in general are characterized by higher rates of impulsivity compared to non-gamblers.[28]

Internet Games

High sensation-seekers can also be drawn to exciting media experiences, even to the point of addiction. Movies and television shows can be thrilling, but they're nothing compared to the immersive and engaging world of Internet gaming. I'm not talking about a round of the latest color matching games for the iPhone. The multi-billion dollar world of computer games has spawned an emerging psychological condition. The American Psychiatric Association gives us a peek into Internet gaming disorder, a research description of individuals who take computer gaming to the extreme.[29]

For these gamers, Internet games are all-encompassing. They stock up on supplies of sports drinks, easily microwaved foods, and adult diapers to wear during gaming binges so they don't have to miss a moment of play. One website devoted to gaming hosted an online confessional where gamers admitted the extreme measures they had taken for gaming binges. One admitted taking a week off from school to indulge in a new game. He even came up with an ingenious way to prevent his sister from knowing he was hiding in his room to play.

"My senior year of college, I decided I was going to take a week off. I also couldn't go to the restroom because the bathroom was right in front of her room. So I had to make do and started peeing in some old coffee cans. I would dump and rinse them out when I 'got home from school.' It was gross, but the game was awesome and it was totally worth it." Another waited for his fiancée to leave town to indulge, but he lost track of time: "I was confused sometime later when she came back in the house and had a concerned look on her face. I hadn't really noticed but I had been playing the game non-stop for 2 days straight and she had come back from her trip. She forced me to take a shower before I passed out."[30]

In extreme cases, Internet gaming can cause problems more severe than a few missed days of class and a stinky room. Some people have lost jobs due to missed work, had children removed from the home because of neglect, and even died due to renal failure.[31]

And guess what? Sensation-seekers are more likely to get addicted to Internet gaming. Jie Wang and his colleagues examined 37 studies of Internet addiction and found a relationship between Internet addiction and sensation-seeking.[32] Among the four components of sensation-seeking, disinhibition was the subscale that was most closely linked to Internet addiction.

Aggression

Just as sensation-seekers can jump right into their pleasure-seeking pursuits, they can also be impulsive in their expressions of aggression. When I was in fifth grade, I was the target of a mysterious expression of aggression. A kid in my class passed me a note that said I should meet him after school so he could beat me up. What's even more perplexing to my adult mind was that I chose to meet him after school. I can only imagine that my fifth grader logic determined that if I didn't show up, things would be worse. I arrived to the melee with a plan. I asked if it was okay to put down my steel sided Six Million Dollar Man lunch box. But instead, as I bent down, with all the force I could muster, I slammed the corner of the lunchbox into his ribs. He doubled over, and I ran as fast as I could for home. Dirty fighting? Maybe. But at the time it seemed to be my best option to avoid a bloody nose.

I never did find out what set that kid off that day, but psychologists know in general what seems to spark aggression and other disruptive behaviors, not only in fifth graders but also in adults. Early theories of anger regarded this emotion as a reaction to a goal blocked or unattained. Known as the frustration-aggression hypothesis, this theory suggests that frustration occurs when a goal is blocked, leading to anger and aggression, which are words or physical acts that a person does to cause harm.[33] While a blocked goal can elicit aggression, it can also produce embarrassment, guilt, and nervousness.

Your peer group can influence how you express your frustration at blocked goals. If your parents ever worried that you might "run with the wrong crowd" as a kid, it turns out they may have been rightly concerned. Who we hang out with can influence disruptive behaviors. We tend to choose friends who are like us, a process called social selection. Fairly inhibited people with relatively low levels of impulsivity may reinforce each other's tendency to think through their choices when things don't go their way. They aren't as inclined to lash out or attack and can come up with alternatives to deal with frustrations. But adolescents with dispositions toward delinquency, for example, tend to select, and be selected into, groups of kids who also have a propensity to delinquency.[34] Once in that group, it tends to increase the likelihood of engaging in those behaviors.

Impulsivity and disinhibition make a powerful combination when people get frustrated. For some sensation-seeking people, frustrations can lead to problems with anger and aggression. In 2013, researchers Jeff Joireman, Jonathan Anderson, and Alan Strathman found that disinhibition and boredom susceptibility were related to anger, hostility, and physical and verbal aggression.[35] Sensation-seekers may be less likely to use avoidance strategies like disengagement or find less risky ways to deal with frustrations. Sensation-seekers who score high on boredom susceptibility and disinhibition may also be attracted to situations where aggression is likely. They may have little concern for the future consequences of their behavior and are more likely to become hostile and angry and engage in aggression in those situations. They also report feeling calmer after fighting. They may not only be more likely to let an altercation escalate into a fight, but also sometimes they go hunting for one.

Of all the forms of self-control, perhaps one of the most important to master has to be the ability to control aggressive behavior. Even the most mild mannered of us must conjure an image of running our shopping cart into a fellow shopper who is wandering aimlessly and slowly as you are trying to make your way efficiently through the shopping aisles (I call them meanderthals). Why don't I ram them with my metal shopping cart? Many reasons, not the least of which is that the immediate consequences are pretty easy to figure out and there are much better ways to achieve your final goal. Acting aggressively is sometimes the first thing people think to do, but most of the time it's not the best option, and many times it's the worst one. Despite this (too) many people engage in aggressive acts despite what must seem like obvious negative consequences.

Using a series of four studies involving 573 women and 272 male college students, Jeff Joireman and his research team investigated the relationships between sensation-seeking, impulsivity, and a focus on immediate consequences of behavior. They used a theory called the General Aggression Model (GAM).[36]

The GAM recognized the complex nature of aggression. It suggests that biology, environment, feelings, thoughts, and how revved up you are can all influence aggressive or non-aggressive behavior.[37] The GAM says that your personality and the situation you are in can influence how you perceive your situation. How you perceive your situation can in turn create emotions which can cause arousal. This arousal can influence how you assess your situation and can influence what decisions you make. All of this together influences if you act aggressively or non-aggressively.

Another twist on the GAM is that Jeff Joireman's team also examined an individual's ability to consider the outcomes of their behavior or CFC (consideration of future consequences). Joireman's group used a 12-item scale with questions such as "I consider how things might be in the future and try to influence those things with my day to day behavior."[38]

What did they find? Sensation-seeking seems to affect aggression because it influences hostile thoughts and anger. What's more, high sensation-seekers are drawn to situations that could elicit aggression. There's more. Scores in disinhibition, in particular, were the best way to predict physical aggression and boredom susceptibility was the best predictor of verbal aggression. For some sensation-seekers aggression is "fun," but the researchers noticed this brings about a paradox.

Some sensation-seekers are more likely to engage in aggression in part because it relieves boredom, but also because they often harbor feelings of hostility and anger. Why? Over time the aggressive behavior that sensation-seekers participate in (you'll remember from the relationship chapter that many high sensation-seekers love to push people's buttons) can produce negative reactions from other people. This could be another reason why high sensation-seekers experience a higher level of hostility and anger than average and low sensation-seekers.

What about consideration of future consequences? Don't high sensation-seekers know that acting aggressively might cause problems? They do, but they might not care. High sensation-seekers are often okay with their decision to act aggressively, at least until the consequences sink in.

Remember Devin from earlier? He told me that aggressive behavior used to be a big part of his life. "I used to get into fights. No reason really, it was a natural high for me. It's a big thing that thrill-seekers get involved in."

Sensation-seeking is not only linked with aggression but also in some cases other kinds of disruptive behaviors. There are links to high levels of sensation-seeking in children and disruptive behaviors such as stealing, destroying property or being suspended from school.[39] For some, these disruptive behaviors might be motivated by boredom or anger. Betty Pfefferbaum and Peter Wood interviewed individuals who had been involved in property destruction. Some 38 percent said they did it for fun, and 11 percent reported that the motivation was anger or revenge.[40]

The bottom line is this: Sensation-seeking *can* lead to dark places if it goes too far. When their behaviors become too extreme or their sensation-seeking becomes the whole focus of their life, HSSs can put themselves and the people around them in danger. In my experience these are edge cases. The vast majority of high sensation-seekers I spoke to were not only extremely friendly, but also very respectful of others and generally had a logic around the risks they took for themselves.

Is there a way to know which high sensation-seekers lean toward the dangerous end of the spectrum? Is it possible to tell in advance who will take their quest for the buzz too far? There may be, and the best test may be a marshmallow.

The Most Famous Marshmallow in the World

Back in the 1960s, psychologist Walter Michelle and his team presented children with a wondrous and terrible challenge.[41] They offered the kids a marshmallow, a gooey, sticky, sweet marshmallow. It's important to remember that to a little kid, a marshmallow is like a miniature cotton candy party and nearly irresistible. Michelle's team offered each of the kids a single marshmallow and a deal. They could eat the marshmallow immediately (sounds good), or if they waited until the experimenter returned, they could have two (doubly good). So, either eat it now and have immediate gratification, or wait, delay gratification, and double your pleasure.

The researcher struck the deal, then left. What happened next was fascinating. Some kids gobbled down the marshmallow even before the researcher stood up to leave. They weren't able to delay their gratification, even though it meant that they wouldn't get a second treat. The other kids waited, realizing that if they could just hold off on what they wanted for a short amount of time, they would be rewarded. In order to delay gratification, they did everything they could to distract themselves. Their distraction behaviors ran the gamut from hilarious to ingenious to nearly too painful to watch. They sang songs, closed their eyes, and even pretended to eat the marshmallow. One of them sniffed the marshmallow so frantically that she started to hyperventilate. Some found other creative ways to deal with the frustration of not getting what they wanted immediately.

The researchers followed up with the kids years later and discovered some important differences between the two groups. The children who had been able to delay gratification were better off in all sorts of ways. For example, they had better academic outcomes and were less likely to have problems with drug addiction.

Delay of gratification is a nifty skill that comes in handy more than you might think. If you are able to delay gratification, then it's easier to tolerate frustration, save money, not yell when you are angry, and resist temptation. If you aren't able to delay gratification, then it's harder not to give in to the things you desire, despite the positive or negative consequences of waiting.

This ability to delay gratification is a lot like disinhibition. People who score high on the sensation-seeking scale of disinhibition have trouble holding back. If they want to do something, they do it. If you are able to delay gratification, it means that you have the ability to control yourself at least some of the time, and delay gratification in the face of temptation, especially when you aren't being watched.

You may have noticed over the course of this chapter that many of the problematic behaviors HSSs end up engaging in are associated with disinhibition and boredom susceptibility. Thrill- and adventure-seeking and experience-seeking can play a role as well, but based on the research, they seem less likely to drive the HSS to the dark side. Michelle's experiment gives us an important clue as to why. If you can't delay gratification, if disinhibition and boredom susceptibility drive you to take more risks, indulge in substances, and behave aggressively, you are setting the stage for problems physically, socially, and otherwise.

Icarus got the warning about the importance of moderation a bit too late. But how can you tell if sensation-seeking has reached a point where it's a problem? As one counselor pointed out to me, "It's not a problem until it's a problem." There are many common characteristics among the various addictive behaviors, and it's good to note if sensation-seeking behavior is becoming a problem. The key is to take note when and if sensation-seeking behavior is getting out of control. Constantly thinking about an activity, engaging in a behavior even though it is causing harm, withdrawal symptoms when not engaging in the behavior, or even hiding the behavior after people have expressed their concern – these are all tell-tale signs the sensation-seeker has taken things too far. The more those characteristics look familiar, the more problematic your behavior might be. If you're on a mountain experiencing headaches, vomiting, and your face is turning blue, but you're considering finishing the climb anyway? That's a sign. If your friends tease you because you have the most dented car in the group and they are terrified to drive with you? That's a sign – even if you're convinced you're the best driver on the planet. And, of course, if you're facing problems at work, school or with other responsibilities, or experiencing social problems caused by obsessive sensation-seeking, that's a strong signal that something needs to change. Consider, too, if you have lost friends, have had frequent injuries, or your friends are concerned about your sensation-seeking behavior. If you feel your behavior is causing a problem, it's important to consult a mental health professional such as a licensed psychologist or licensed professional counselor for help.

Addiction is never healthy. Aggression is rarely useful. And risk *is* risky. In its most extreme forms, sensation-seeking can lead to addiction, injury, and death. If you let your sensation-seeking drive you too close to the edge, you may just fall off. However, despite the fact that it definitely has a dark side and it remains a mystery to me in many ways, I have developed a great deal of respect for the high sensation-seeking personality. In fact, I see it as a kind of superpower as you'll learn in the next chapter.

8 SUPER POWER OR SUPER PROBLEM

When I first began my journey to understand high sensation-seekers, I honestly wondered what was wrong with them. People who threw themselves off of buildings and out of airplanes, chose the most unusual thing on the menu, and would pick a topic of conversation that they knew would cause conflict? Who does that?

Could Freud be right? Maybe Thanatos was stronger in high sensation-seekers and created an unruly, pandemonium-loving personality. Was sensation-seeking some kind of death wish? Were high sensation-seekers simply lackadaisical about the beauty and fragility of life? Were they *actually* chaos junkies or adrenaline addicts? Or was high sensation-seeking a neurological impairment that makes people want to BASE jump or rollerblade downhill in city traffic or try ever more challenging, even poisonous exotic foods? I was baffled. It seemed, at the time, that what they really sought was chaos.

For a low sensation-seeker like me, initially it was difficult to intuitively understand the point of the activities that high sensation-seekers engaged in with such gusto. I even had more than a few of my psychologist colleagues suggest that perhaps past trauma contributed to their current "rash" behavior. However, after reading hundreds of pages of research, logging hundreds more hours interviewing high sensation-seekers, and observing them engage in their seemingly outlandish activities, a more complex picture began to come into focus.

Like pretty much every other personality trait, high sensation-seeking can help or it can get in the way. And like pretty much any personality trait, it all depends on how it's folded into your life and how extreme the behavior becomes. Take extroversion. We don't think about extroversion as a problem to be fixed. In fact, extroversion is encouraged and celebrated in our culture. However, if you take extroversion to an extreme you end up with an overpowered extrovert with few boundaries and so eager to fill their social interaction tank they annoy the people around them.

It's the same with sensation-seeking, just because someone is a high sensation-seeker, doesn't mean they are irresponsible and out of control. In fact, for some, it's precisely the opposite. It is their deep love of life and the almost desperate knowledge that it is fleeting that drives them to live close to the bone and "suck the marrow out of life" as Thoreau once put it.[1] I have friends who when trying a new food might take a small bite to see if they like it. They nibble at the smallest piece to get the tiniest taste of it. My high sensation-seeking friends are different. They'll chomp like hungry goats filling their mouths with an unfamiliar food and decide to accept the gustatory consequences once it hits their taste buds. Just like with food, high sensation-seekers take the biggest bite out of life they can. They want to savor and intensify their experiences, perhaps because they are deeply aware this moment, right now, may be their one shot at doing something extraordinary, and they are serious about collecting those experiences for the "museum of the mind" that Victor described in Chapter 3.

At the same time, it's clear that in some cases the behaviors just go too far, and high sensation-seekers who become too extreme can harm themselves and others. Stalling a plane over the Gulf of Mexico and expecting your new date to pull you out of it isn't safe. Driving too fast and too close, because you feel like you can is dangerous. And, as you have learned, high sensation-seekers have an increased risk of drug use and addiction, go through divorces more often, and sometimes will take their behaviors to the very edge of death.

With that said, I have rarely met a group of people so friendly, so generous with their time, or so willing to explain and share their experiences. Indeed, most of the high sensation-seekers I met wanted, more than anything, for me to understand them, even to be one of them. They wanted me to feel the rush that comes when you are in free fall during a skydive, or the explosion of sensation you experience when you eat a new unexplainable flavor for the very first time, or the calm, Zen-like, focus that comes when you scuba dive in rivers with bull sharks and alligators during storms. Like someone offering a slice of their grandmother's beloved German chocolate cake, they nearly pleaded with me to try the things they love to do. "Try it, just TRY IT!"

The thing is, I can't. I can't experience these activities as high sensation-seekers do. I'm not physically or psychologically wired for it. Even if I did try, I couldn't experience it through a high sensation-seeker's lens. It would just be a hodgepodge of panic and alarm for me (mostly panic).

This realization is part of what has made me appreciate that high sensation-seeking is a really special trait. It can be like a superpower, or it can turn into a super problem. When it's used for good, it can be beneficial, even healing and protective. But like with most superpowers, it also has a dark side and when sensation-seeking runs amuck, it can be used for ill and even destroy the sensation-seekers' lives and those around them.

To close, I want to explore this final realization and explain why, after much reflection, I think high sensation-seeking is closer to being a superpower than a super problem and I want to explore how it is used for good by the high sensation-seekers around us every day.

Living without Fear

One of the most prominent characteristics of high sensation-seekers – especially those with high scores on thrill- and adventure-seeking and disinhibition – is their seeming lack of fear. Most of us would never consider jumping from the Perrine Bridge in Idaho, because our natural survival instinct would kick in. The average person also wouldn't intentionally eat deadly pufferfish or blurt out whatever we want in public just to see the reaction our words provoke in others. Anxiety – whether it's social anxiety or the very real threat of death – keeps our behavior in check.

Indeed, fear has always been essential to our survival. Without it, we probably wouldn't have made it very far. The fight or flight response protects us from danger both real and imagined. Fear is so essential, so utterly primeval that it is almost hard-wired into one of the deepest areas of our nervous system, the amygdala.

The amygdala is a bit of brain matter in the temporal lobe. Neurons in the amygdala are linked to emotions such as fear. Fear is not only essential for humans, but also a crucial response in all animal life. If you want to survive, avoiding things that could end your existence is a good place to start.

However, since high sensation-seekers have a different response to highly chaotic and sometimes dangerous experiences, they don't always perceive they're in danger when they are, in fact, in danger. Take Lara, a high sensation-seeker who lives in California. I asked her about her early experiences with thrill.

"I think the earliest I can remember is being about five," she explained. "My dad had this old '53 Ford pickup and we would go drive after church. We'd drive up into the mountains in Northern California. I would say, 'Stop! Stop! You've got to stop the truck!' My dad would pull over and then I'd get out and I'd stand on the edge of the cliff and hold my arms out and say, 'This is so beautiful!'

"My mom would scream 'Lara, get back in the car! Get back in the car!' And my dad would get out and just stand next to me and put his hand out in front of me so I didn't fall over the edge. I loved it. I wasn't afraid. I don't know if that makes me a thrill-seeker, but I didn't feel any fear. I think they probably thought that stuff would go away, but it didn't."

And boy didn't it. In fact, Lara's sensation-seeking became even more intense as she grew older.

"I've sky dived a few times. I've bungee jumped a couple of times too. Those were probably in my twenties and thirties. Roller coasters, all that. I want the biggest ones I can find. Sometimes I go on them by myself because my family won't go with me."

I asked her if she'd ever felt afraid.

"Of dying? No, I don't think so."

"I almost hit a train once," she told me as an afterthought. "It was early in the morning, I was driving and the sun was shining behind me, right into the railroad track lights, you know, the red blinking lights, and I couldn't see them. There were no safety rails that came down. I've driven over that same train track a thousand times. It looked like the sun was shining and I just proceeded to go through it like I always would. About the same time, I was about to cross the railroad tracks the train was suddenly there.

"I locked eyes with the conductor. He looked at me and his face said 'You're about to die.' It was the fear in his eye. I believed it. There was fear in his eyes and I felt the fear. I hit the brakes and slid. My car spun sideways. When I finally came to a stop the train continued on and I could have reached out my window and touched the train I was so close to it. I think I sat there for probably 20 minutes, then I put the car into gear and drove on."

"Where did you go?" I asked her.
"Oh . . . I went to the mall."

Most people in a situation like this would have been panicked. Lara's ability to stay calm in a chaotic situation helped her to avoid the train plowing into her car and to go calmly about her day afterwards. There is a reason people say they are "paralyzed with fear." As we learned in Chapter 2, the fight, flee, or freeze response leads some people to freeze in fear. This would have prevented Lara from acting.

After listening to Lara's story, I wondered what a life without fear might be like. I mean who really *wants* to be afraid? I came across an episode of one of my favorite podcasts, *Invisibilia,* that happened to focus on fear. According to their website, *Invisibilia* "explores the intangible forces that shape human behaviors. Things like ideas, beliefs, assumptions and emotions." The program's name is Latin for "all the invisible things." The fear episode features a woman (they refer to her as SM to protect her identity) who suffers from Urbach-Wiethe.[2]

Urbach-Wiethe, also known as lipoid proteinosis or hyalinosis cutis, is a genetic condition characterized by both skin and neurological lesions. It's incredibly rare with fewer than 400 reported cases since it was first discovered by two medical doctors (Erich Urbach and Camilo Wiethe) in the late 1920s.[3] Because of a mutation of a particular chromosome there is a buildup of hyaline (a transparent protein) in the body. The result is a variety of symptoms including dry and scarred skin, eyelid papules (little raised bumps on the skin), a hoarse voice, scarring of the skin, and in some cases, it can result in hardening of brain tissue as well. For SM, Urbach Wiethe resulted in her amygdala becoming calcified. The result is that SM is incapable of feeling fear.

In a rare interview,[4] SM describes in an extremely hoarse, childlike voice, experiences that would cause profound horror in most of us. Yet, she relates them in a shockingly matter-of-fact way.

"Years ago," she starts, "when my three sons were small, I was walking to the store, and I saw this man on a park bench. He said, 'Come here please.' So I went over to him. I said, 'What do you need?' He grabbed me by the shirt, and he held a knife to my throat and told me he was going to cut me. I said, 'Go ahead and cut me. I'll be coming back, and I'll hunt your ass.' I wasn't afraid. And for some reason, he let me go. And I went home."

She sounds brave. But is that really bravery? Being brave means you persist *despite* the fear you feel. I mean you aren't really brave if you don't know you are in danger. Because SM didn't experience fear she was unable to detect she was in danger, so she dealt with the situation with a certain direct logic: If you attempt to harm me, I will track you down and kill you.

Apparently, SM has had encounters like this more than once. She's been held at knife point multiple times, held at gun point twice, and her ex-husband almost beat her to death. Louise Spiegel, the journalist who wrote the story, speculates that because SM has no fear she is more open to human interaction both good and bad. Most of us can sniff out dangerous situations, and we try to avoid them. But SM doesn't seem to have any dowsing rod for danger. On the one hand, this makes her about the friendliest person you could hope to meet. On the other, it sets her up to be exploited by the seedier elements of society.

You might think she would be traumatized by so many attacks, except one outcome of her total lack of fear is that she also doesn't experience trauma. Trauma occurs, in part, when a person has difficulty making sense of a shocking or disturbing experience. People sometimes try to cope with trauma by numbing themselves, so they can't experience the pain. Often people who experience traumas relive the experience in their minds over and over again and they are on the lookout for new sources of trauma all the time, to avoid being harmed again. This can manifest itself in what psychologists call hypervigilance or a heightened sensitivity to your environment. Without fear you cannot be traumatized. So when SM relates these stories, she isn't thinking back to a terrible time or day in her life. It's just a fact that it happened like the time the wrong package was delivered to your house. That's it.

Scientists have conducted experiments on SM to demonstrate her lack of fear. Experiments that honestly seem a little gruesome. They have put her in a room with dangerous snakes and other animals and had to physically restrain her from playing with them. They even attempted to condition a fear response by exposing her to loud horns and other harsh sounds – but even that didn't work. In one of the studies she was asked to draw a face with a fearful expression. She wasn't even able to conjure an image of a fearful face in her mind.

SM reports actually *enjoying* the extreme situations in her life. Instead of running away, she turns to face them. Being held at knife point wasn't a bad thing for her, because she didn't experience the horror the rest of us would. In fact, she describes herself as quite happy, "9 out of 10," she says.

It sounds familiar doesn't it? Don't get me wrong, high sensation-seekers obviously *do* have fear. That said, we also know that high sensation-seekers respond to fear differently than most of us do all the way down to the neurochemical level. They are able to remain calm in extreme circumstances that would send most of us running, and it affects not only the way they perceive those experiences but also the impact the experiences have on them as well.

Sensation-Seeking and Trauma

Dover is a major ferry port in south-east England. Across the Strait of Dover is the port of Zeebrugge, a village on the coast of Belgium. It's just 84 nautical miles from Zeebrugge to Dover. The trip takes approximately 3 hours by passenger and car ferry. People drive their cars on to the ferry so they can use them once they arrive in the next city. It's a typical roll-on/roll-off passenger ferry. However, on March 6, 1987, the roll-on/roll-off passenger ferry *Herald of Free Enterprise* left Zeebrugge carrying 459 passengers and 80 crew. It's suspected that someone left the bow door open and when the ferry left port, water rushed into it. It flooded and capsized immediately. A total of 193 people died in the frigid waters.[5] As you can imagine it was an incredibly traumatic experience.

Five years later, researchers Stephen Joseph, Chris Brewin, William Yule, and Ruth Williams found 35 survivors of the disaster and had all of them complete a series of psychological questionnaires.[6] The researchers discovered that those with high PTSD symptoms scored higher on impulsiveness than those low in PTSD symptoms.

What seems to happen with some people who survive traumatic events is a greater pursuit of risk. As a consequence, they look for dangerous situations to explore. Because of this relationship, "reckless and self-destructive behavior" has been added as a symptom of PTSD in the fifth edition of the *Diagnostic and Statistical Manual of Mental Disorders*, the handbook used to diagnose psychiatric conditions.[7]

One study, for example, involved over 200 veterans diagnosed with PTSD.[8] The researchers also assessed the amount and severity of reckless behaviors. Reckless and self-destructive behavior was reported by over 74 percent of the participants with 61 percent engaging in multiple forms. These behaviors included dangerous drug use, alcohol use, operating a car while impaired, or engaging in aggressive behavior such as fighting. In fact, the more symptoms of PTSD a person had, the more of these behaviors they reported. "These types of high-risk behaviors appear to be common among veterans who have experienced trauma, and put veterans in harm's way by making it more likely that they will experience stress and adversity in the future," says Naomi Sadeh of the National Center for PTSD at the VA Boston Health Care system.[9]

Because of this, many mental health professionals ask me if people with high sensation-seeking personalities are suffering from trauma. It's easy to link Freud's idea of Thanatos (the death instinct) with survivors' guilt from traumatic experiences. If that's what is happening you would assume that people with PTSD symptoms would also have greater sensation-seeking tendencies, except many studies point to the opposite. For example, Scott Orr and his colleagues found that war veterans with PTSD, although having greater boredom susceptibility than those in the control group, actually had lower scores on sensation-seeking.[10]

What's more, most people who are high sensation-seekers have had that personality trait for as long as they can remember. They have memories even as kids of doing high sensation-seeking stunts. If a person is suddenly impulsive and reckless after a traumatic experience it may be linked with PTSD and might not be part of a high sensation-seeking personality.

Analysis is Paralysis: Acting without Fear

One of the first people I interviewed for this book was a friend of mine named Andrew. Andrew is an archetypal high sensation-seeker. You remember him from the early chapters. He was in the military, went to the police academy, became a deputy sheriff, and finally settled in the field of counseling psychology all because he has a deep desire to help people. But what stands out about Andrew the most is his ability to selflessly jump into dangerous circumstances to help others – a behavior for which he has something of a reputation.

Once we were walking along in my neighborhood with a group of people during a neighborhood festival. It was a beautiful day and there were people everywhere. A car drove by with a huge dog who had half his body sticking out the car window taking in the wind. Then something happened. The dog leapt out of the moving car and, uninjured, tore down the street after a squirrel. Was the dog okay? Where was it going? The people in the car pulled over. Everyone was shocked, except for Andrew. I looked around and he, too, was gone. Without missing a beat, he raced down the street through the crowds, was able to nab the dog, and reunited him with his owners. All of this before any of us realized what was going on.

Here's another example of Andrew's altruistic nature. "It was just after I left the Sheriff's Department," Andrew explained. "I was riding with a friend in San Francisco. We were talking about being a cop and then were stopped at a traffic light behind some other cars. We were near this alleyway and we both look down there, and there's this guy threatening a woman who is pregnant. But we're just watching this guy threatening and holding this woman against a wall. I jumped out of the car and ran toward them and confronted them. I broke up the fight. When I got back to the car my friend said, 'I turned around and you were gone. They could have had a knife. They could have had a gun. You've got to think about those things.' I actually think it's important that we don't think about those things and we just act."

One of the things that struck me in this story was that last line: Just act. That's something I heard over and over again from high sensation-seekers. In fact, some of them would use the phrase "analysis is paralysis." Instead of analyzing situations they jump headlong into danger and trust their bodies and minds to respond as needed in the present moment. The goal is not to think about what to do too soon and of course not too late.

All in all, that's not such a bad thing. Many of us do the opposite and analyze each worry that pops into our mind. There's a classic study that demonstrates this. Psychological researchers had people keep a daily log of all the things that worried them throughout the day. "I'll run out of gas on the way to work." "I'll forget my friend's birthday next week." "They'll run out of bananas at the grocery store." You get the idea. They captured the rambling flow of worry that many of us subject ourselves to each day. Then the researchers followed up. Did they get to work on time, remember their friend's birthday, did they have bananas? A stunning 85 percent of the concerns never came to pass. And for the concerns that did, the participants reported that they handled them much better than they anticipated.[11] Nevertheless, most of us get stuck in our negative ruminations, a sort of recreational anxiety and self-doubt of our ability to handle what life might throw our way.

High sensation-seekers are different. During their high sensation activities most handle the task in the moment without too much analysis. The rest of us tend to overanalyze situations that never arise and not act on the ones that do. Not such a great combo. We'd like to think we would act if the situation demanded it. Instead many of us stand passively by hoping someone else will. This is called "the bystander effect." It has been demonstrated repeatedly that in dangerous situations – in situations where someone is getting hurt or worse – the more people watching a crime, the less likely any of them are to intervene. This may be because we assume other bystanders will act, or it may be because we are paralyzed by fear. It takes a person like Andrew to break the chain of passivity. Perhaps if we weren't so trapped in our anxious, worrying minds, the story would be different. For at least part of the population, it certainly seems to be. Can acting without fear – even without thought in some cases – be problematic? Of course, it can. But I think we need people like this around when things really go south.

The Dark Knight

High sensation-seekers are often our protectors. Anton, an army ranger who did search and rescue missions to recover downed pilots in the military, was the first person who related a metaphor to me that I came to understand is broadly known in the military and law enforcement communities. We will call it "the sheep, the wolves, and the sheepdog." As far as I know it was first written about in a book called *On Combat* by Dave Grossman in which he describes the psychology of war and deadly conflict.[12] The metaphor was related to him by a retired colonel who was a Viet Nam veteran. Here is the basic premise.

Most people are decent and not capable of hurting each other. These people are the sheep. We probably don't like to think of ourselves as sheep, but I suspect there might be an element of truth in this. And the vet who told the story says he means nothing negative by calling them sheep, "To me it is like the pretty, blue robin's egg. Inside it is soft and gooey but someday it will grow into something wonderful. But the egg cannot survive without its hard blue shell. Police officers, soldiers and other warriors are like that shell, and someday the civilization they protect will grow into something wonderful."

A much smaller percentage of the population has a capacity for violence. People who have this capacity and little to no empathy for others become vicious criminals. These are the wolves. But what about those who have a capacity for violence and a deep love for others? These are the sheepdogs, and their job is to protect the sheep from the wolves. Some are military or police officers who use their capacity for aggression to protect others.

This idea can, of course, be a slippery slope. It's also very easy to excuse violence by justifying it as protection. But whatever you feel about the police and the military or your personal beliefs on the politics of this metaphor, it resonates very deeply with soldiers, police officers, and some first responders. Anton is a perfect example. When he left the military he got a job at the VA specifically to help address this issue. "One of the reasons I went to counseling school and one of the reasons I do what I do in my private practice is because when I came back from the war, every one of the counselors they had at the VA, was in my mind, a sheep. They're trying to treat me like I was a sheep and I was like, 'I'm a sheepdog. I don't need my hypervigilance as you call it, to go away, because that's the skill I have that keeps me alive when I'm in combat.'" And for some sheepdogs they are constantly in combat. On the lookout for the wolves among us.

Interestingly, Anton is now helping vets do just that. "A lot of what I do is help them [returning vets] define what their flock looks like. If you can't be in the military anymore, what can you do to still feed that inner protector that you have, and then what can you do to still feed that need for some sort of action, some sort of activity that challenges your skillset."

What kinds of flocks do these people find? What activities challenge their skill set? As you might imagine, it ranges. Some go into social work and enter child protective services. Others participate and coach Spartan Races. Still others, like Anton, counsel their peers realizing that the sheepdogs themselves are the flock they most want to help and protect. The good news is that, with Anton's help, many are able to find a way back into the world.

If you think about it, it's too bad that we need military and police for protection; but we sometimes do. And I wish I could say that everyone with authority uses that authority wisely; they don't. And it's not that every person who comes back from war or joins the police force is a sheepdog, and they certainly aren't all high sensation-seekers. However, high sensation-seekers who are able to be hypervigilant and experience potentially anguishing experiences without the same effects much of the rest of us would suffer through, that's a super power. Now, if they can also take that ability and help protect society in an ethical, connected way, that's even better.

The Healers and the Helpers

Many high sensation-seekers have a deep desire to help others. Andrew and Anton are both former military individuals who became mental health professionals. I actually ran into this somewhat surprising transition far more often than I expected to while on this path to understanding HSSs. I guess I originally thought HSSs would be selfish. Maybe I believed this because high sensation-seekers seemed hedonistic and pleasure loving; or possibly it was my misguided perception that they were mostly driven by their pursuit of the buzz. What I found was generally the opposite: The vast majority of the people I talked to were somehow engaged in helping others.

The ER doctors and nurses you met in Chapter 6 dedicate their lives to saving people. While we learned that being a high sensation-seeker was a part of the puzzle for them, it was indeed only a piece of it. Timmy O'Neil, the philosopher of thrill you met in Chapter 4, founded a company called Paradox Sports (which you will learn more about in a moment) to help people with physical disabilities engage in the same kinds of extreme sports he so enjoys. Jason, the underwater Indiana Jones from Chapter 6, is virtually addicted to making new discoveries to help further our understanding of the natural world. Even Devin, the gentleman with a history of drug and alcohol abuse from the last chapter, ultimately turned his addictions around and now counsels those with similar problems.

Could it be that high sensation-seekers also seek to help others? A study done in 2008 at Auburn University paints an interesting picture of the relationship between high sensation-seekers and the desire to help others.[13] In the study, scientists interviewed 1,100 college students who were extreme sports enthusiasts to find out how motivated they were to engage in civic participation. The outcome? While high sensation-seeking did not correlate with current civic participation, it *was* associated with the desire to work for activist and reform-oriented organizations, specifically in leadership roles.

Not too surprising. This is what high sensation-seekers do. It's who they are. These are the people who are out at the edges pushing the bounds of human experience, trying to reshape the world around them. They do this out on the hillside, they do it in the boardroom, they do it for other people. And sometimes they do it for all three.

But can fearlessness also come with a dark side? Let's think about the attributes of what makes a good hero. Impulsiveness so she can jump in and help at a moment's notice, boldness, and at times aggressiveness, can be helpful too – for example, to help resolve tough conflicts. It's also most likely the case that the same list of attributes can be applied to villains as well. Some researchers believe that many of the ingredients of what makes a person a psychopath are really the same ingredients that can make a person a hero. David Lykken suggested that the two are "twigs off the same branch."[14]

In fact, these same attributes could be helpful for politics, business, the military, or even extreme sports. Boldness, self-focus, and a defiant personality were linked with both altruism and a concept the researchers called everyday heroism. Everyday heroism (sometimes called small "h" heroism) are the little helpful things that people do every day. Big "H" heroism involves big risks like death or injury, but small h heroism are things like helping others, or showing kindness. Christina Patton, Sara Francis Smith, and Scott Lilienfeld surveyed 251 first responders and 170 who didn't have that job.[15] The researchers discovered that no matter their profession, fearlessness, dominance, boldness and, yes, sensation-seeking were all correlated with everyday heroism. That means that the higher your sensation-seeking score, the more likely you are to be involved in those acts of everyday heroism. It takes a certain personality for both big and small h heroism. Frank

Farley, the L.H. Carnell Professor at Temple University, says "Heroes . . . often have difficulty articulating why they did it, often saying, 'I simply had to do it' or 'I just did it, I didn't think about it.'"[16] That fits with how some high sensation-seekers just act and trust that things will just work out. But there's more to the equation, according to Farley: "However, I believe two ingredients are included in most Big H heroism, and those are risk-taking/risk-tolerance and generosity (compassion, kindness, altruism, empathy). If an individual is significantly risk averse he or she is very unlikely to place their life at risk for the sake of someone else. They might want to do it for a host of good reasons, but they simply cannot do it."[17]

Paradox Sports: Making the Impossible Possible

I want to close with one of the most touching stories I encountered in writing this book. It revolves around a company called Paradox Sports, which was founded by Timmy O'Neil who you met in Chapter 4. You want to know what the "paradox" is in Paradox Sports? They develop hardware and services to help people who need adaptive climbing assistance – most often amputees and people with limb differences (many of whom are coming home from war) – to participate in extreme sports.[18] These are high sensation-seekers who are helping their own flock – other high sensation-seekers – reengage in the sensation-seeking activities they love after injuries most would assume would make it impossible.

What does this look like? Imagine 18 amputees in Ouray, Colorado, doing some of the most extreme ice climbing imaginable. Or a blind woman snowboarding, not just down the junior slopes mind you, but executing and landing jumps that would make most of us weak in the knees. Or a high school student who couldn't walk, taking his first steps after downhill skiing in a chair.

These are the kind of people Mike Neustedter, the executive director of Paradox Sports, has been working with all of his life.

The blind woman on the snowboard – that is Mike's best friend from college. After she lost the use of her eyes, Mike was determined to help her snowboard again, because she had begun to believe her plans for the future vanished with her sight. "I wanted to find a sport or something that would get her involved in something that she couldn't do. That was the easiest way for me to provide her with the power to feel that even though she has a disability she can do everything that she originally planned on doing before her disability." The outcome? She is now married to a fellow snowboarding enthusiast, she has two sons, and they ski the slopes around Tahoe all the time. All of these were dreams she thought were behind her once she lost her sight.

The group of amputees on Ouray – that was the first expedition Mike conducted with Paradox. "We develop different ice climbing attachments so people who have an amputation or a limb difference, they are able to participate in the sport as well. The incredible thing about ice climbing is, as far as people with disabilities, there's not much difference between an able-bodied climber and an adaptive climber. It looks the same. Same tools, same equipment that any able-bodied person would use."

And the guy who started walking after taking up skiing with Mike? His name is Marshall Garber, and his story is perhaps the most touching of all. When he was a freshman in high school, a tumor was found on Marshall's spine. It had to be removed, even though the surgery meant there was a chance he would never walk again. Post-surgery, the doctor's fears were realized and Marshall was paralyzed from the waist down.

After several years of intense therapy, Marshall was barely able to stand up. That's when he met Mike and the Paradox Sports team. Marshall came with a group from Kennedy Krieger Medical Center to do some skiing. He was sit-skiing (where you sit in a chair instead of standing up), and was having a great time. Every day, after his skiing, doctors would run tests on Marshall to see what impact this kind of program could have on patients' lives. The results spoke for themselves.

During the five-day program Marshall was able to stand up from his wheelchair. Doctors were dumbfounded. No one really knew what was going on. One theory was that the skiing may have activated feelings in his spine that either hadn't been there before or he hadn't experienced in a long time.

Resilience and hardiness are personality styles exhibited by people who can persevere despite the adversity they experience. Hardiness and resilience act as buffers against stress. People with these qualities have lower stress levels, fewer illnesses,[19] and greater job satisfaction.[20] Samuel McKay, Jason Skues, and Ben Williams of the Swinburne University of Technology in Victoria, Australia, had 268 people complete questionnaires that measured sensation-seeking, stress and trauma, coping, perceived reliance and wellbeing.[21] They found that the higher the level of sensation-seeking, the greater the life satisfaction and resilience, and the lower the perceived stress.

How does sensation-seeking create resilience? High sensation-seekers see potential stressors as challenges to be overcome rather than threats that might crush them. Rather than dodging or running from a threat, they dive into it head on (analysis is paralysis, after all). Because of this they are more likely to successfully meet challenges. This not only helps them to resolve challenges, but also boosts their belief in their ability to handle such challenges in the future. It's a buffer against the stress of life. It's this mindset that increases their hardiness and resilience in the long term. Several studies have supported this idea, and researchers such as McKay, Skues, and Williams have found (not surprisingly) that high sensation-seekers have lower amounts of information overload, are impacted less by daily stressors (such as losing your keys, things breaking around the house or feeling like you have too many things to do) and have lower discomfort for the stressors they do experience. What's more, when they experience a significant setback from a stressor, they bounce back more quickly. Why? High sensation-seekers tend to engage in more problem-focused approaches to the stressor, meaning they target the cause of stress in a practical way. High sensation-seekers have lower perceived stress, greater positive emotions, and report greater life satisfaction.

Despite how much I crave predictability, life is often far from predictable. And when unwanted negative surprises pop up, sometimes I panic. Novelty for some can be a burden. High sensation-seekers, on the other hand, bask in the novelty for the positive payoff it often involves. Sensation-seekers see novelty and surprise as less teetering, less overwhelming, and easier to control. What's more, when they look back, they are happy with how they handled these surprises as compared to low and average sensation-seekers who are often not.

Hardiness may have been the factor that helped Marshall Garber to keep going. Mike agrees, "I think what it gave him was the mental 'I can do it' attitude. That allowed him to stand. He was always able to stand out of his wheelchair, but it wasn't until he had that confidence and the drive that actually gave his body the power to do it."

Needless to say, everyone was thrilled with what happened. But the incredible story doesn't end there. Three months after the program ended, Mike got an email from Marshall. He was able to stand at his high school graduation – which had been a personal dream for him – and he continues to do sit-skiing and sit-biking. But the most surprising part of all: He is now able to walk about 100 feet with a walker.

Mike's response to all of this is not only interesting, but also the whole reason he's involved with Paradox. "How much longer would he have been in that frame of mind that he couldn't stand up if it wasn't for these action sports. Being able to do it gave him the motivation to do it, and that took his therapy to the next level. All of these sports, we consider them therapeutic adventure activities, and I think that therapeutic word is key."

Can action sports be therapeutic? Could being a high sensation-seeker actually have protective and restorative powers? Is it possible that adventure skiing is what gave Marshall the hardiness, will, and fortitude needed to take his therapy to the next level? It's hard to know for sure, but what I have discovered is that there is a connection between high sensation-seekers, their chosen activities, and their mental and physical health.

Do you remember Wes, the cyanotic mountain climber from the last chapter? I caught up with him a little over a year after our first interview. He had some bad news.

"It was December 27th, I was heading home at night around 7:30, 7:45. I'm driving, and then, all of a sudden, I see a flash – something coming toward me. It was a guy who was more than twice the legal limit, and he was driving on the wrong side of the road. He came at me head on. His lights weren't fully on. He hit me just head on. I have two herniated discs in my neck, some joint issues in my neck, some nerve damage in my foot, some back issues and stuff like that."

Because of his accident, his doctor said he couldn't lift more than 20 pounds and for quite some time he was unable to engage in the physical activities he so loved. His dreams of being a fire fighter are shattered. He used to do Taekwondo, not anymore. No more CrossFit. He suffered a deep depression as a result. But things are changing for him and he knows his limits. "I think I'm much more conservative. You won't see me doing any skydiving, you won't see me doing any BASE jumping or stuff like that. But I still want to live my life, you know what I mean?" Once he did, his mood started to improve. He's shifted from thrill- and adventure-seeking and has embraced his experience-seeking side. "It's the best way to refresh myself. I bought a Lonely Planet book. I'm buying a new camera. I leave tomorrow for Japan, Taiwan and Hong Kong. It's like, getting lost might not be a bad thing."

These are all extreme examples of another thread of this story we have now seen over and over: Extreme activities bring high sensation-seekers peace, confidence, and happiness. They might even have a positive impact on their physical biology.

How to Stay in the Light

We've examined some of the potential woes and wonders of high sensation-seeking. You might wonder how to tip the scales and get more good than bad from this personality trait. Here are a few recommendations for doing just that.

First ... stop and think. High sensation-seekers are known for their ability to act and have their body figure out what to do. It's often one of the things they are most proud of doing. However, every now and then it's good to stop and think to decide if the activity or action being considered is really worth the risk.

Next, separate sensation-seeking from novelty seeking. Occasionally, when people feel bored, what they're looking for is something new. The newness doesn't HAVE to come from thrill- and adventure-seeking. Novelty can come without the risk. The newness, for example, could come from doing something routine in a very different way.

Also, increase empathy. Earlier, we learned that sometimes high sensation-seekers may have trouble seeing things from another point of view. The more that a high sensation-seeker can increase their empathy for others, the more it might help these relationships. In fact, I think increasing empathy can be helpful for everyone.

So, how can you increase empathy? In his article, "The Six Habits of Highly Empathic People", Roman Krznaric (pronounced *kriz-NAR-ik*) gives a few tips to increase empathy.[22] For instance, cultivate your curiosity about strangers, challenge your prejudices, discover commonalities between yourself and others, or even try another person's life in your imagination.

Next, try reducing disinhibition and boredom susceptibility. The two parts of the sensation-seeking personality that are likely to get a high sensation-seeker in trouble are disinhibition and boredom susceptibility. How can you reduce these? There is some evidence that mindfulness mediation can help. Meditation can be a powerful tool to fight off boredom. I know that meditation may not seem like the most exciting activity, especially to a high sensation-seeker, but it can be a big benefit to both high and low sensation-seekers alike. When you start meditating, start small. Even two minutes is a great place to begin. Do a little research on some of the benefits of mindfulness meditation. I think you'll be surprised. And do it your own way. Meditation doesn't have to mean sitting in a quiet room for hours. Instead, try a walking meditation.

My final suggestion for helping to be sure that sensation-seeking helps more than it hurts has to do with finding what sensation-seeking activities are best for you. High sensation-seekers do best when they promote healthy activities like adventure sports. If you don't have a lot of friends who are high sensation-seekers, consider sharing experience-seeking activities with your average and low sensation-seeking friends.

These tips, which are actually good tips for all of us, can help to make sure high sensation-seekers are getting the best from their superpower.

CONCLUSION

Where does all of this leave us? Is high sensation-seeking a superpower, as illustrated in the last chapter, or is it a super peril, as discussed earlier? What of the questions I set out to answer in this book? Are high sensation-seekers different than the rest of us? Is there something wrong with them? Is being a high sensation-seeker dangerous? Should we try to change their behavior?

Obviously, it's complicated

In retrospect, I think I may have been asking the wrong questions. "Is there something wrong with them?" Well, no. "Are these people different than the rest of us?" Again, no. And yes. They are also different from each other. "Is being a high sensation-seeker dangerous?" It can be. Bungee jumping off a bridge is dangerous. Hanging off the side of a building with one hand? Dangerous. Eating fugu? You got it, dangerous. BASE jumping? Not as safe as sitting on the beach, I'd guess. But all of us engage in dangerous activities all the time. Cliché as it sounds, every time you get in your car or on a plane, or cross the street, you take your life in your own hands, don't you? Life is dangerous. Perhaps with the proper precautions and perspective skydiving is not as life threatening as it looks to the rest of us.

High sensation-seeking does have risks. We have seen that it can lead to addictive behaviors that harm people and relationships. It can also lead to a kind of impulsivity that would strike most of us as nearly pathological. Of course, these behaviors also occur in the rest of the population, likely in about the same percentages as in HSSs. And high sensation-seeking can also be a wonderful gift.

For example, one of the key unifying themes I found throughout my research is that HSSs have a tendency to engage in daily activities that provide a sense of awe, that goose-bump laden feeling that we all know. I have started to think of them as awe-seekers. Think about the kinds of things we have seen sensation-seekers doing: whether it's racing around town at 100 miles an hour, experiencing numb lips from fugu, running obstacle courses, BASE jumping, or even eating pig's blood stew, experiencing awe is part of the reward. It turns out that awe is pretty good for your body, and this may be one of the reasons HSSs go out of their way to seek it.

Researchers from UC Berkeley asked 94 students to fill out questionnaires that told the researchers how frequently the students experienced different emotions. The students supplied saliva samples, which were then analyzed for interleukin-6 (IL-6), a molecule known to promote inflammation throughout the body. Inflammation is tied to poor health, and low IL-6 might signal good health. Happy emotions were linked to lower IL-6 levels. But the strongest correlation was with a surprising emotion: awe. The more frequently someone reported having felt awestruck, the lower the IL-6.[1]

Could it be that the HSS's physiology is encouraging them to seek out awe? Do they drive fast, even recklessly at times, to drive down inflammation levels? Could eating goat cheese-flavored ice cream or a bowl of chicken hearts actually make someone healthier? Is it possible that doing a Tough Mudder race may improve your health instead of endangering it?

These questions are beyond the scope of this book, and, in truth, we may never have the answers to them. But one thing is clear: high sensation-seekers spend much of their daily lives pushing the bill to seek out experiences that promote a sense of awe. When it comes to the human condition, few emotions are as powerful or as telling about who we are and what we value. In a world that is increasingly mechanized, atomized, and polarized, perhaps what the high sensation-seeker is really after is the same thing the ancient mystics were seeking – a sense of wonder about the world in which we live. Is this a super power? I'm not sure. But personally, I can't think of a better buzz.

Here are my conclusions. High sensation-seeking is like any other personality trait or any other tool. It can be positive or negative depending on how you use it. In the end (and I'm shocked to admit this), I think the good probably outweighs the bad. High sensation-seekers do incredible things – things most of us would never dream of doing – and they love it. That's not such a bad way to live your life.

It's also not a way I want to live my life. Lots of HSSs tried to convince me I should be a high sensation-seeker too. "Don't you want to know what it feels like?" Not really. I'm not an HSS. It's not my thing. You can keep the crazy stunts, radical social interactions, and even the wild food. It's not for me. I have no desire at all to bungee jump or do extreme snowboarding or race cars or engage in virtually any of the other behaviors you have read about in this book. Zero. None. Not at all interested. Not even in the slightest.

But what if you aren't a high sensation-seeker? Is there something we can learn from them? Yes! A lot. Here are three that stand out.

1. Go with the flow. Sensation-seekers are often trying to achieve a "flow state." We know that flow is healthy for us. It helps us to enjoy life, and be cheerful, satisfied, creative, and have higher self-esteem. It also fends off stress. Flow also enhances learning, which could come in handy for thrill-seekers and pretty much everyone else. But you don't have to leap from tall buildings to engage in flow. There are plenty of average and even low sensation-seeking experiences that will help you to envelop in your own flow-inducing activities including playing or listening to music, engaging in physical activities, or gardening.
2. Feel the awe. We know that high sensation-seekers collect awe inducing experiences. They know what creates awe for them and they are experts at gathering those experiences. They drain interleukin-6 and fill their minds, body, and sprit with that awe. If scaling buildings gives you more anxiety than awe, you can find that sense of awe in activities that are closer to the ground. For me it's art, nature, and sunrises.
3. Try new things. Not liking something isn't the worst thing that can happen to you. When some average and

> low sensation-seekers are introduced to something new and unusual, they often reject it simply because it is unfamiliar or unusual. It might be good to ask yourself, "What's the worst thing that could happen if I try this?" We might be missing out on new and wonderful experiences just for the fear of not liking it.

So, while you may not take up Olympic skeleton racing anytime soon, learning about sensation-seekers may inspire you to expand your experiences a little more. As the father of flow Csikszentmihalyi suggests "The best moments in our lives are not the passive, receptive relaxing times . . . the best moments usually occur if a person's body or mind is stretched to its limits in a voluntary effort to accomplish something difficult and worthwhile."[2]

I think high sensation-seekers could afford to have a little more empathy for those of us who don't have the nervous system or environmental conditioning that creates the appetite for danger, adventure, or novelty that they do. We low sensation-seekers are your counterparts. You need us as anchors, as lookouts to prevent too much danger and risk. If you think about it, since we feel more risk and fear and tolerate it, we are the brave ones.

With that said, I think the opposite is also true. Those of us on the lower sensation-seeking end of things could afford to stop ogling at HSSs and assuming they are crazy thrill-seekers with a death wish and no regard for life. I found this to be an almost universally incorrect assumption. Like the rest of us, high sensation-seekers are complex, dynamic, puzzling, fascinating people who find meaning and purpose in what they love to do – even if what they love to do does kinda, sometimes look a little nuts to the rest of us. I'm a geeky academic who prefers sunrises to flying through breathtaking vistas in a squirrel suit. But I'm also a firm believer in the incredible biological and social complexities that make those experiences possible and even desirable for people who find it thrilling. It would be a sadly diminished world if there were no high sensation-seekers in it. However, I will continue to look on in awe at the extraordinary things they do from a comfortable, safe distance.

APPENDIX 1

For Students: Learning Objectives and Topics for Discussion

By Jonna Kwiatokowski

Chapter 1

Learning Objectives

- Differentiate between typical behaviors and traits of high sensation-seekers versus low sensation-seekers.
- Use personality theories to understand the traits that might be confused with sensation-seeking.
- Understand how Marvin Zuckerman identified sensation-seeking as a unique trait.
- Explore a measure of sensation-seeking that includes subscales for thrill- and adventure-seeking, experience-seeking, disinhibition, and boredom susceptibility.
- Interpret your sensation-seeking score and compare it to other possible profiles of scores.

Discussion Questions

1. Based on the various personality theories, what are some myths about sensation-seekers that are debunked? Were you surprised by any of the personality traits that are not necessarily related to sensation-seeking?
2. Imagine yourself in Zuckerman's Ganzfeld Procedure, or try it with these simple instructions: www.instructables.com/id/Ganzfeld-Hack-Your-Brain-the-Legal-Way/. How do you imagine your response would compare to those Zuckerman eventually labeled as high sensation-seekers?
3. Think of a favorite high thrill-seeking fictional character. What profile would you assign to that character

across the four subscales: 1) thrill- and adventure-seeking, 2) experience-seeking, 3) disinhibition, and 4) boredom susceptibility.
4. Think of a person who is difficult for you to understand or relate to. To the best of your knowledge, how is that person's profile of sensation-seeking traits the same and different from your own? How might differences in sensation-seeking preferences challenge professional and personal relationships?
5. What is your sensation-seeking profile? Is it surprising to you? How does it match or fail to match your job requirements?

Chapter 2

Learning Objectives

- Understand that optimal experiences differ by individual, and that these differences can be linked to genetic influences.
- Compare the evolutionary benefits of being a low versus a high sensation-seeker as they are represented by the Behavioral Avoidance and Behavioral Inhibition Systems.
- Interpret sensation-seeking behaviors that are found in studies of human testosterone and MAO (monoamine oxidase), and studies of rat stress hormones.
- Revisit the four subtypes of sensation-seekers from a biological perspective.
- Consider non-biological factors that might influence sensation-seeking, including how you might influence gene expression through methylation.

Discussion Questions

1. The Goldilocks zone is a repeated concept through this chapter. What is the "just right" formula for a sensation-seeker, in terms of both biological and social factors?
2. It's easy to see the evolutionary advantages and disadvantages to sensation-seeking for humans from our distant past. Do those advantages and disadvantages translate to modern human challenges?
3. There are many psychotropic drug options that influence a person's testosterone or MAO levels. What are

the implications of taking these drugs toward sensation-seeking behaviors?

4. The examples in this chapter might lead one to think that high sensation-seekers simply do not experience stress, or at the very least that stressors have a different effect on them. Thinking of a high sensation-seeker that you know, is this the case? How is stress experienced the same and differently by high sensation-seekers?
5. Methylation suggests that you might have the ability to influence your children's genes toward more or less sensation-seeking behaviors. Assuming you had that control, would you act in ways that would encourage sensation-seeking in your children?

Chapter 3

Learning Objectives

- Explore how sensation-seekers experience cultural opportunities such as comedy, music, and art.
- Understand how high sensation-seekers use multitasking, and might use it to sustain their optimal arousal level.
- Notice how travel, adventure, and authentic experiences overlap for the high sensation-seeker.
- Notice how food variety, hotness, unusualness, danger, and cultural curiosity might drive a high sensation-seeker.

Discussion Questions

1. Multitasking is presented as a preferred work pattern for sensation-seekers. This is possibly due to the boost in stimulation one gets when switching to a new task. Might this also relate to optimal level theory? Can multitasking help sensation-seekers sustain a needed level of arousal?
2. If we know high sensation-seekers feel a benefit from multitasking and we know that multitasking has inherent dangers, is there a safer way that sensation-seekers can keep boredom at bay without overloading their senses with multiple tasks?
3. Victor is an example in this chapter of someone who collects experiences in the "museum of his mind." Do

you have a mind museum? What do you put in your mind museum? Can you have a non-adventure-based mind museum?

4. What is the hottest food you have ever consumed? Most unusual? Dangerous? What drove you to try these foods? Did you enjoy the experiences?

Chapter 4

Learning Objectives

- Recognize the role of arousal potential, the ability of an activity to draw one's attention, in the choice of activities for high sensation-seekers.
- Balance the role of arousal potential with the importance of experiencing greater focus and self-control through risky activities.
- Understand grit, a "passion and perseverance for a long-term goal," as a contributing factor in the development of the skills required for the complex and risky activities undertaken by high sensation-seekers.
- Appreciate that sensation-seekers are not necessarily trying to induce a high from stress (through adrenaline), but instead want to manage and tamper with stress through skill.
- Analyze the role of social media in encouraging high sensation-seekers.
- Consider the implications of improving the life skills of agency and emotion regulation through high-risk activities.

Discussion Questions

1. The cases described in this chapter seem focused on some of the less tangible benefits of risk-taking. High sensation-seekers want complex problems that cannot be solved directly, or that might present unexpected challenges throughout. What are the life skills honed by high sensation-seekers beyond simply taking a risk?
2. The author suggests that high sensation-seekers may have grit, which is described as "sticking with your future, day in day out, not just for the week, not just for the month but for years and working really hard to make that future a reality." Would a high sensation-seeker

sustain interest in learning skills that take so much time, given that sensation-seekers also report that they get bored easily? Are there any clues in Katherine Beatlie's account of learning to backflip her wheelchair?
3. At least two of the people (Will and Jeb) in this chapter state that they dislike adrenaline, which is often cited as a main benefit from people addicted to running (i.e., the "runner's high"). Think about runners or other people you know who claim to be "adrenaline junkies" and compare them to profiles of high sensation-seekers. Is there overlap? Are people who seek an adrenaline rush from more "regular" activities more similar to low sensation-seekers?
4. Social media encouragement is cited as desirable to high sensation-seekers. Do you think it is a primary motivating factor for a high sensation-seeker, or a secondary factor to the risky activity itself?
5. Some of the high sensation-seekers in this chapter seem to be quite philosophical about their risk-taking behaviors. Do you think their philosophical inquiries prompted sensation-seeking in the first place, or is that an after-effect of death-defying experiences?
6. Is it possible that sensation-seeking and meditation lead to the same outcomes? Where might that comparison sustain, and where might it fold?

Chapter 5

Learning Objectives

- Compare how high sensation-seekers and low sensation-seekers prefer to interact with other people.
- Understand that high sensation-seekers extend their risk-taking behaviors into interpersonal relationships by disclosing more to casual and close friends, engaging in disagreement, and encouraging others to join them in similar behaviors.
- Recognize that high sensation-seekers may seek relationships for similar reasons to why they seek adventurous activities. They may be less interested in lasting bonds or deep connections than a moderate to low sensation-seeker.

- Consider that the high sensation-seekers' typical pattern of engaging in relationships may not always lead to desirable outcomes, but odds may improve when high sensation-seekers find each other.

Discussion Questions

1. Search the website "The Ultimate Guide to Worldwide Etiquette" (www.swissotel.com/promo/etiquette-map/) and choose a country that you have never visited. Are there behavioral norms for this country that are surprising to you? Would any of these adjustments in your behavior make you feel uncomfortable? Why or why not?
2. Imagine a high sensation-seeker engaged in a romantic relationship with someone seeking each of the six styles of love (agape, eros, storge, pragma, mania, and ludos). Predict the trajectory and ultimate outcome for each relationship.
3. A common complaint about people who organize their social lives through social media is that they are always looking for something better to do, and may engage in activities briefly before searching for the next best thing, and jumping to that. Has social media created or encouraged more sensation-seeking behaviors in the general population? Are more of us choosing to "keep our options open"?
4. High sensation-seekers show lower emotional intelligence, and in particular they tend to show low interpersonal intelligence. At the same time, we know from earlier chapters that high sensation-seekers are drawn to complex problem solving and love a challenge. What might make interpersonal interactions a less interesting challenge for a high sensation-seeker? (It is certainly plenty complex!)

Chapter 6

Learning Objectives

- Compare various professions that might attract high sensation-seekers and notice how the job requirements draw different subtypes of sensation-seekers.
- Recognize that some jobs require high sensation-seeking for success, but most do not.

- Consider how a high sensation-seeker justifies risky behavior in a work situation. They will often identify external challenges that caused them to take a risk.
- Reflect on how a high sensation-seeker might approach choosing a career. Jobs might support high sensation-seeking directly through the inherent risks or indirectly by providing time or funds for risky non-work-related activities.

Discussion Questions

1. Given the choice between doing something that you know you can do or something that you aren't sure you can do, which would you choose? What aspects of the job or task, such as the potential to help others or learn something new, might change your answer?
2. Everyone has to take risks sometimes and everyone experiences success and failure sometimes. Think of a time when a risk that you took led to a failure. How did you think about that experience at the time? How did you see your role in the failure? What external factors did you identify as contributing to the situation?
3. Consider the jobs that you have had in your life and the activities that you do for pleasure. Is there coherence between your work life and non-work life?
4. This chapter presents evidence suggesting that high sensation-seeking CEOs might be more successful. Would you want to work for a high sensation-seeker?

Chapter 7

Learning Objectives

- Consider the "zero risk model" to explain the risk-taking patterns of a high sensation-seeker. It suggests that high sensation-seekers do not see their behaviors as overly risky, but that they may modify their choices if they are presented with evidence that they are taking too big of a risk.
- Understand that high sensation-seekers are particularly susceptible to a "time-saving bias," possibly as another external justification for risky behaviors such as driving too fast or following other cars too closely.

- Review the typical stages that someone might go through from curiosity about drugs to addiction to drugs, especially for high sensation-seekers.
- Connect aggression and other disruptive behaviors as forms of risky behavior that can be satisfying to the high sensation-seeker.

Discussion Questions

1. Like Icarus, we can all sometimes "fly too high" or "fly too low." How do you know when you are at your extremes in a situation, and when you are moderating to appropriate levels? What is your comfort zone for risk?
2. If you don't value sensation-seeking, what is it in your life that you value highly or even the most? What kind of risks would you take for those things you value highly?
3. The author lists a host of risky driving behaviors that have been associated with high sensation-seekers. Should our communities do more to identify high sensation-seeking drivers? Should there be additional training or information sessions for sensation-seekers who want to drive on the roads?
4. Based on what you've reviewed in this book, how would you define and recognize healthy versus unhealthy versions of sensation-seeking?
5. Conduct your own version of the marshmallow test with a friend. Identify a highly satisfying food or drink that you both enjoy and set it between you while you talk. How long can you wait before indulging? Is there a difference between you in terms of how long you can delay gratification? Does the pattern relate to your relative propensity for sensation-seeking?

Chapter 8

Learning Objectives

- Develop an appreciation for the unique combination of characteristics, maybe even superpowers, you are likely to find in high sensation-seekers.
- Understand the implication of living without a strong sense of fear, with no fear. It allows for quick action

in the face of danger, quick recovery from dangerous situations, calmer evaluation during dangerous situations, and possibly even some level of enjoyment from danger.

- Differentiate recklessness, which is a common behavior after a traumatic event, from sensation-seeking, which is not typically related to trauma survivors.
- Consider that high sensation-seekers are often also the helpers and healers in our communities. The same set of characteristics that urge them to take risks for themselves may also encourage them to take risks for you.
- Appreciate that the many risks taken by high sensation-seekers may build a resilience through direct knowledge of their limits and abilities.

Discussion Questions

1. Remember a time when you have felt fear in a dangerous situation, maybe like Lara felt when she almost hit a train. During the danger, what was your response? What was your response after the danger had passed?
2. Consider all of the potential "superpowers" of the high sensation-seeker. Which ones do you see in yourself? Which ones are less evident?
3. As you review the characteristics of a sensation-seeker, do you agree with the author that it is more of a superpower than a super problem?
4. Has this book influenced the way you interact with people, be they high, medium or low sensation-seekers? Is there added understanding of what drives risk-taking behaviors? Or, maybe of how subtypes of sensation-seeking can drive different choice patterns?
5. Is it possible to glean sensation-seeking styles from early interactions with people? If so, has this book influenced how you will seek out friend, romantic, or professional relationships with people based on their sensation-seeking profile?

APPENDIX 2

Zuckerman's Sensation-Seeking Scale Form V

Directions: Each of the items below contains two choices. Please indicate which of the choices most describes your likes or the way you feel. There are no right or wrong answers. For each question, give yourself 1 point for A and 0 points for B.

Thrill- and Adventure-Seeking

1	A. I often wish I could be a mountain climber.	
	B. I can't understand people who risk their necks climbing mountains.	
2	A. I sometimes like to do things that are a little frightening.	
	B. A sensible person avoids activities that are dangerous.	
3	A. I would like to take up water skiing.	
	B. I would not like to take up water skiing.	
4	A. I would like to try surfing.	
	B. I would not like to try surfing.	
5	A. I would like to learn to fly an airplane.	
	B. I would not like to learn to fly an airplane.	
6	A. I would like to go scuba diving.	
	B. I prefer the surface of the water rather than being under water.	

(*cont.*)

7	A. I would like to try skydiving.	
	B. I would never want to try jumping out of a plane, with or without a parachute.	
8	A. I like to dive off the high dive board.	
	B. I don't like the feeling I get standing on the high dive board (or I don't go near it at all).	
9	A. I would like to sail a long distance in a small but seaworthy sailing craft.	
	B. Sailing long distances in small sailing crafts is foolhardy.	
10	A. I think I would enjoy the sensations of skiing very fast down a high mountain slope.	
	B. Skiing down a high mountain slope is a good way to end up on crutches.	

_____ **Thrill- and Adventure-Seeking Score**

Experience-Seeking

1	A. I like some earthly body smells.	
	B. I dislike all body odors.	
2	A. I like to explore a strange city or section of town by myself, even if it means getting lost.	
	B. I prefer a guide when I am in a place I don't know well.	
3	A. I have tried marijuana for fun or would like to.	
	B. I would never smoke marijuana for fun.	
4	A. I would like to try drugs that produce hallucinations.	
	B. I would not like to try any drug that might produce strange effects in me.	

(*cont.*)

5	A. I like to try new foods that I have never tasted before.	
	B. I order foods with which I am familiar, so as to avoid disappointment and unpleasantness.	
6	A. I would like to take off on a trip with no preplanned or definite routes or timetable.	
	B. When I go on a trip, I like to plan my route and timetable fairly carefully.	
7	A. I would like to make friends who are unconventional.	
	B. I prefer the "down to earth" kinds of people as friends.	
8	A. I like to meet people who are very different from me.	
	B. I tend to stay away from people who are different from me.	
9.	A. I often find beauty in the "clashing" colors and irregular forms of modern paintings.	
	B. The essence of good art is in its clarity, symmetry of form and harmony of colors.	
10	A. People should dress in individual ways even if the effects are sometimes strange.	
	B. People should dress according to some standard of taste, neatness, and style.	

_____ **Experience-Seeking Score**

Disinhibition

1	A. I like "wild," uninhibited parties.	
	B. I prefer quiet parties with good conversation.	

(*cont.*)

2	A. I don't mind people who are uninhibited about sex.	
	B. I dislike people who are uninhibited about sex.	
3	A. I often like to get high (drinking alcohol or smoking marijuana).	
	B. I find that stimulants make me uncomfortable.	
4	A. I like to have new and exciting experiences and sensations, even if they are a little frightening, unconventional, or illegal.	
	B. I am not interested in experience for its own sake.	
5	A. I am attracted to people who are physically exciting.	
	B. I like to date people who share my values.	
6	A. Keeping drinks full is the key to a good party.	
	B. Heavy drinking usually ruins a party because some people get loud and boisterous.	
7	A. A person should have considerable sexual experience before marriage.	
	B. It's better if two married persons begin their sexual experience with each other.	
8	A. I would love to live the lifestyle of the super-wealthy.	
	B. Even if I had the money, I would not care to associate with other wealthy people.	
9	A. I enjoy watching many of the sexy scenes in movies.	
	B. There is altogether too much portrayal of sex in movies.	
10	A. I feel best after having a couple of drinks.	
	B. Something is wrong with people who need alcohol to feel good.	

_____ **Disinhibition Score**

Boredom Susceptibility

1	A. I can't stand watching a movie that I've seen before.	
	B. There are some movies I enjoy seeing a second or even a third time.	
2	A. I get bored seeing the same old faces.	
	B. I like the comfort and familiarity of everyday friends.	
3	A. When you can predict almost everything a person will do and say, he or she must be boring.	
	B. I dislike people who do or say things just to shock or upset others.	
4	A. I usually don't enjoy a movie or play when I can predict what will happen in advance.	
	B. I don't mind watching a movie or a play when I can predict what will happen in advance.	
5	A. Looking at someone's personal videos or travel photos bores me tremendously.	
	B. I enjoy looking at personal videos or travel photos.	
6	A. I prefer friends who are excitingly unpredictable.	
	B. I prefer friends who are reliable and predictable.	
7	A. I get very restless if I have to stay home for any length of time.	
	B. I enjoy spending time in the familiar surroundings of home.	
8	A. The worst social sin is to be a bore.	
	B. The worst social sin is to be rude.	
9	A. I like people who are sharp and witty, even if they do sometimes insult others.	

(*cont.*)

	B. I dislike people who have their fun at the expense of hurting the feelings of others.	
10	A. I have no patience with dull or boring people.	
	B. I find something interesting in almost every person I talk to.	

____ **Boredom Susceptibility Score**

Total Sensation-Seeking ____

This questionnaire is provided for general information only, and should not be treated as a substitute for the medical advice of your own health care professional. Always consult your own health practitioner if you're in any way concerned about your health.

You can interpret your score using this scale:

Score Profile
0–16 Low Sensation-Seeking
17–27 Medium Sensation-Seeking
28–40 High Sensation-Seeking

For each component
0–3 Low
4–6 Medium
7–10 High

Sensation-seeking consists of four distinct components

- Thrill- and adventure-seeking – quest for risk
- Experience-seeking – love of new sensations of the mind and senses
- Disinhibition – ability to be unrestrained
- Boredom susceptibility – dislike for repetition

The sensation-seeking survey reveals five scores: one for each component of sensation-seeking and a total sensation-seeking score. For each of the individual components of sensation-seeking a score of 0–3 is low, 4–6 is average, and 7–10 would be high. For total scores 0–16 would mean you are a Low Sensation-Seeker, 17–27 is Average Sensation-Seeking and 28–40 would mean you are a High Sensation-Seeker.

These ranges are simple comparisons to the general population. Don't get too worried if your numbers are low or high. Think of the score as a way to understand the relative nature of sensation-seeking. There are no right or wrong scores or best way to be. The scores are just a clue to your individual pattern of seeking or avoiding sensations.

Adapted from Zuckerman, M. (1994).
Behavioral expressions and biosocial bases of sensation seeking.
Cambridge: Cambridge University Press

APPENDIX 3

The AISS (Arnett Inventory of Sensation-Seeking)

For each item, indicate which response best applies to you:

A)	describes me very well
B)	describes me somewhat
C)	does not describe me very well
D)	does not describe me at all

1. I can see how it would be interesting to marry someone from a foreign country.

A)	describes me very well	
B)	describes me somewhat	
C)	does not describe me very well	
D)	does not describe me at all	

2. When the water is very cold, I prefer not to swim even if it is a hot day. (-)

A)	describes me very well	
B)	describes me somewhat	
C)	does not describe me very well	
D)	does not describe me at all	

3. If I have to wait in a long line, I'm usually patient about it. (-)

A)	describes me very well	
B)	describes me somewhat	
C)	does not describe me very well	
D)	does not describe me at all	

4. When I listen to music, I like it to be loud.

A)	describes me very well	
B)	describes me somewhat	
C)	does not describe me very well	
D)	does not describe me at all	

5. When taking a trip, I think it is best to make as few plans as possible and just take it as it comes.

A)	describes me very well	
B)	describes me somewhat	
C)	does not describe me very well	
D)	does not describe me at all	

6. I stay away from movies that are said to be frightening or highly suspenseful. (-)

A)	describes me very well	
B)	describes me somewhat	
C)	does not describe me very well	
D)	does not describe me at all	

7. I think it's fun and exciting to perform or speak before a group.

A)	describes me very well	
B)	describes me somewhat	
C)	does not describe me very well	
D)	does not describe me at all	

8. If I were to go to an amusement park, I would prefer to ride the roller coaster or other fast rides.

A)	describes me very well	
B)	describes me somewhat	
C)	does not describe me very well	
D)	does not describe me at all	

9. I would like to travel to places that are strange and far away.

A)	describes me very well	
B)	describes me somewhat	
C)	does not describe me very well	
D)	does not describe me at all	

10. I would never like to gamble with money, even if I could afford it. (-)

A)	describes me very well	
B)	describes me somewhat	
C)	does not describe me very well	
D)	does not describe me at all	

11. I would have enjoyed being one of the first explorers of an unknown land.

A)	describes me very well	
B)	describes me somewhat	
C)	does not describe me very well	
D)	does not describe me at all	

12. I like a movie where there are a lot of explosions and car chases.

A)	describes me very well	
B)	describes me somewhat	
C)	does not describe me very well	
D)	does not describe me at all	

13. I don't like extremely hot and spicy foods. (-)

A)	describes me very well	
B)	describes me somewhat	
C)	does not describe me very well	
D)	does not describe me at all	

14. In general, I work better when I'm under pressure.

A)	describes me very well	
B)	describes me somewhat	
C)	does not describe me very well	
D)	does not describe me at all	

15. I often like to have the radio or TV on while I'm doing something else, such as reading or cleaning up.

A)	describes me very well	
B)	describes me somewhat	
C)	does not describe me very well	
D)	does not describe me at all	

16. It would be interesting to see a car accident happen.

A)	describes me very well	
B)	describes me somewhat	
C)	does not describe me very well	
D)	does not describe me at all	

17. I think it's best to order something familiar when eating in a restaurant. (-)

A)	describes me very well	
B)	describes me somewhat	
C)	does not describe me very well	
D)	does not describe me at all	

18. I like the feeling of standing next to the edge on a high place and looking down.

A)	describes me very well	
B)	describes me somewhat	
C)	does not describe me very well	
D)	does not describe me at all	

19. If it were possible to visit another planet or the moon for free, I would be among the first in line to sign up.

A)	describes me very well	
B)	describes me somewhat	
C)	does not describe me very well	
D)	does not describe me at all	

20. I can see how it must be exciting to be in a battle during a war.

A)	describes me very well	
B)	describes me somewhat	
C)	does not describe me very well	
D)	does not describe me at all	

Scoring Guide: Combine responses to items, with A = 4, B = 3, C = 2, D = 1, so that higher score = higher sensation-seeking. For items followed by (-), scoring should be *reversed*. A = 1, B = 2, C = 3, D = 4

Novelty subscale

1.
3. (-)
5.
7.
9.
11.
13. (-)
15.
17. (-)
19.

Intensity subscale

2. (-)
4.
6. (-)

8.
10. (-)
12.
14.
16.
18.
20.

Adapted from: Arnett, J. (1994). Sensation seeking: A new conceptualization and a new scale. *Personality and Individual Differences, 16*(2), 289–296

APPENDIX 4

Grit Scale

Please respond to the following 12 items. Be honest – there are no right or wrong answers!

1. I have overcome setbacks to conquer an important challenge.

A)	Very much like me	
B)	Mostly like me	
C)	Somewhat like me	
D)	Not much like me	
E)	Not like me at all	

2. New ideas and projects sometimes distract me from previous ones.*

A)	Very much like me	
B)	Mostly like me	
C)	Somewhat like me	
D)	Not much like me	
E)	Not like me at all	

3. My interests change from year to year.*

A)	Very much like me	
B)	Mostly like me	
C)	Somewhat like me	
D)	Not much like me	
E)	Not like me at all	

4. Setbacks do not discourage me.

A)	Very much like me	
B)	Mostly like me	
C)	Somewhat like me	
D)	Not much like me	
E)	Not like me at all	

5. I have been obsessed with a certain idea or project for a short time but later lost interest.*

A)	Very much like me	
B)	Mostly like me	
C)	Somewhat like me	
D)	Not much like me	
E)	Not like me at all	

6. I am a hard worker.

A)	Very much like me	
B)	Mostly like me	
C)	Somewhat like me	
D)	Not much like me	
E)	Not like me at all	

7. I often set a goal but later choose to pursue a different one.*

A)	Very much like me	
B)	Mostly like me	
C)	Somewhat like me	
D)	Not much like me	
E)	Not like me at all	

8. I have difficulty maintaining my focus on projects that take more than a few months to complete.*

A)	Very much like me	
B)	Mostly like me	
C)	Somewhat like me	
D)	Not much like me	
E)	Not like me at all	

9. I finish what I begin.

A)	Very much like me	
B)	Mostly like me	
C)	Somewhat like me	
D)	Not much like me	
E)	Not like me at all	

10. I have achieved a goal that took years of work.

A)	Very much like me	
B)	Mostly like me	
C)	Somewhat like me	
D)	Not much like me	
E)	Not like me at all	

11. I become interested in new pursuits every few months.*

A)	Very much like me	
B)	Mostly like me	
C)	Somewhat like me	
D)	Not much like me	
E)	Not like me at all	

12. I am diligent.

A)	Very much like me	
B)	Mostly like me	
C)	Somewhat like me	
D)	Not much like me	
E)	Not like me at all	

Grit Scale Scoring

Step 1: For questions with **1, 4, 6, 9, 10,** and **12**, assign the following points:

5 = Very much like me
4 = Mostly like me
3 = Somewhat like me
2 = Not much like me
1 = Not like me at all

Step 2: For questions **2, 3, 5, 7, 8,** and **11**, assign the following points:

1 = Very much like me
2 = Mostly like me
3 = Somewhat like me
4 = Not much like me
5 = Not like me at all

Step 3: Add up all the points and divide by 12.

Grit Score: ________

What does my score mean?

- The maximum score on this scale is 5 for extremely gritty.
- The lowest score on this scale is 1 for not at all gritty.

What is Grit?

- Grit is defined as perseverance and passion for long-term goals.

- It entails working strenuously toward challenges, maintaining effort and interest over years despite failure, adversity, and plateaus in progress.
- Grit is unrelated to talent and can be built through a growth mindset.

Duckworth, A.L., Peterson, C., Matthews, M.D., & Kelly, D.R. (2007). Grit: Perseverance and passion for long-term goals. *Journal of Personality and Social Psychology, 9,* 1087–1101.

INVENTORY OF NOTES

Chapter 1: What is Sensation-Seeking?

1. For privacy reasons some names and identifying information (such as age and city) have been changed throughout the book.
2. Take a look at her blog and website www.challengesophie.com/
3. Take a look at his Instagram www.instagram.com/kirbase/?hl=en
4. O'Flynn, K. (2004). Meet the Russian kids who take the world's riskiest photos. *Rolling Stone*, May 6. www.rollingstone.com/culture/culture-news/high-times-meet-the-russian-kids-who-take-the-worlds-riskiest-photos-100170/
5. Ockerman, E. (2016). Pamplona running of the bulls: its surprising history. *Time*, July, 6. www.time.com/4386999/pamplona-spain-running-of-the-bulls/
6. Wills, M., and Wills, C.A. (1981). *Manbirds: Hang gliders & hang gliding*. Englewood Cliffs, NJ: Prentice-Hall.
7. Freud, S. (1964) [1933]. *Why war*. In *The standard edition of the complete works of Sigmund Freud*, Vol. XXII (pp. 179–218). London: Hogarth Press.
8. Jung, C. (1971). *Psychological types*. London: Routledge.
9. Zuckerman, M. (1994). *Behavioral expressions and biosocial bases of sensation seeking*. Cambridge: Cambridge University Press.
10. Zuckerman, M., Kuhlman, D.M., Joireman, J., Teta, P., and Kraft, M. (1993). A comparison of three structural models for personality: The Big Three, the Big Five, and the Alternative Five. *Journal of Personality and Social Psychology*, 65(4), 757–768. https://doi.org/10.1037/0022-3514.65.4.757
11. Eysenck, H.J., and Eysenck, J.H. (1975). *Manual of the Eysenck personality questionnaire*. London: Hodder and Stoughton.
12. Geen, R.G., McCown, E.J., and Broyles, J.W. (1985). Effects of noise on sensitivity of introverts and extraverts to signals in a vigilance task. *Personality and Individual Differences*, 6 (2), 237–241.
13. Eysenck and Eysenck, Manual of the Eysenck personality questionnaire.
14. Zuckerman, Behavioral expressions and biosocial bases of sensation seeking.

15. Costa, P., and McCrae, R.R. (1999). A five-factor theory of personality. *Handbook of Personality: Theory and Research*, *2*, 139–153.
16. Zuckerman, M. (1979). *Sensation seeking: Beyond the optimal level of arousal*. Hillsdale, NJ: Lawrence Erlbaum Associates.
17. Tyson, P.J., Jones, D., and Elcock, J. (2011). *Psychology in social context: Issues and debates*. West Sussex, UK: John Wiley & Sons.
18. Zuckerman, Behavioral expressions and biosocial bases of sensation seeking, 27.
19. Ibid.
20. Arnett, J. (1994). Sensation seeking: A new conceptualization and a new scale. *Personality and Individual Differences*, *16*(2), 289–296. http://doi.org/10.1016/0191-8869(94)90165-1; Zuckerman, M., Kolin, E.A., Price, L., and Zoob, I. (1964). Development of a sensation-seeking scale. *Journal of Consulting Psychology*, *28*(6), 477.
21. Hoyle, R.H., Stephenson, M.T., Palmgreen, P., Lorch, E.P., and Donohew, R.L. (2002). Reliability and validity of a brief measure of sensation seeking. *Personality and Individual Differences*, *32*(3), 401–414. http://doi.org/10.1016/S0191-8869(01)00032-0
22. Radcliffe, S. (2016). Why I ran the London Marathon in body paint. *Challenge Sophie*, April. www.challengesophie.com/blog/category/why-i-ran-the-london-marathon-in-body-paint
23. Zuckerman, M., and Neeb, M. (1979). Sensation seeking and psychopathology. *Psychiatry Research*, *1*(3), 255–264.
24. Cross, C.P., Cyrenne, D.-L.M., and Brown, G.R. (2013). Sex differences in sensation-seeking: A meta-analysis. *Scientific Reports*, *3*, 2486. http://doi.org/10.1038/srep02486

Chapter 2: Born to be Wild

1. Gierland, J. (1996). Go with the flow. *Wired*, 4(9), 160.
2. Csikszentmihalyi, M. (1999). If we are so rich, why aren't we happy? *American Psychologist*, *54*, 821–827. doi: http://dx.doi.org/10.1037/0003-066X.54.10.821
3. Csikszentmihalyi, M. (1988). The flow experience and its significance for human psychology. In M. Csikszentmihalyi and I. Csikszentmihalyi (Eds.), *Optimal experience: Psychological studies of flow in consciousness* (pp. 15–35). Cambridge: Cambridge University Press.

4. Csikszentmihalyi, M. (1998). *Finding flow. The psychology of engagement with everyday life*. New York, NY: Basic Books.
5. Rossin, D., Ro, Y.K., Klein, B.D., and Guo, Y.M. (2009). The effects of flow on learning outcomes in an online information management course. *Journal of Information Systems Education. 20*(1), 87–98.
6. Zuckerman, M. (1979). *Sensation seeking: Beyond the optimal level of arousal*. Hillsdale, NJ: Lawrence Erlbaum Associates.
7. Wundt, W. (1895). Grundzuge der physiologischen Psychologie. *The British Journal of Psychiatry*, *41*(173), 347–348. http://doi.org/10.1192/bjp.41.173.347
8. Zuckerman, Sensation seeking: Beyond the optimal level of arousal.
9. Zuckerman, M., and Neeb, M. (1980). Demographic influences in sensation seeking and expressions of sensation seeking in religion, smoking and driving habits. *Personality and Individual Differences*, *1*(3), 197–206. doi:10.1016/0191-8869(80)90051-3
10. Segal, N.L. (2010). Twins: The finest natural experiment. *Personality and Individual Differences*, *49(4)*, 317–323. doi:10.1016/j.paid.2009.11.014
11. Bouchard, T.J., Lykken, D.T., McGue, M., Segal, N.L., and Tellegen, A. (1990). Sources of human psychological differences: The Minnesota study of twins reared apart. *Science*, *250* (4978), 223–228. doi: 10.1126/science.2218526
12. Hur, Y.M., and Bouchard, T.J. (1997). The genetic correlation between impulsivity and sensation seeking traits. *Behavior Genetics*, *27*(5), 455–463. doi: 10.1023/A:1025674417078
13. Fulker, D.W., Eysenck, S.B. G., and Zuckerman, M. (1980). A genetic and environmental analysis of sensation seeking. *Journal of Research in Personality*, *14*(2), 261–281. doi: 10.1016/0092-6566(80)90033-1
14. Nosowitz, D. (2017). Do not eat, touch, or even inhale the air around the Manchineel Tree. *Atlas Obscura*, March. www.atlasobscura.com/articles/whatever-you-do-do-not-eat-touch-or-even-inhale-the-air-around-the-manchineel-tree
15. Amodio, D.M., Master, S.L., Yee, C.M., and Taylor, S.E. (2008). Neurocognitive components of the behavioral inhibition and activation systems: Implications for theories of self-regulation. *Psychophysiology*, *45*, 11–19. doi: 10.1111/j.1469-8986.2007.00609.x
16. Zarrouf, F.A., Artz, S., Griffith, J., Sirbu, C., and Kommor, M. (2009). Testosterone and depression: systematic review and

meta-analysis. *Journal of Psychiatric Practice*, *15*(4), 289–305. doi: 10.1097/01.pra.0000358315.88931.fc
17. Booth, A., Johnson, D.R., and Granger, D.A. (1999). Testosterone and men's health. *Journal of Behavioral Medicine*, *22*, 1–19. doi: 10.1023/A:1018705001117
18. Dabbs, J.M., Jurkovic, G.J., and Frady, R.L. (1991). Salivary testosterone and cortisol among late adolescent male offenders. *Journal of Abnormal Child Psychology*, *19*(4), 469–478. doi: 10.1007/BF00919089; Yu, Y.Z., and Shi, J.X. (2009). Relationship between levels of testosterone and cortisol in saliva and aggressive behaviors of adolescents. *Biomedical and Environmental Sciences*, *22*(1), 44–49. doi: 10.1016/S0895-3988(09)60021-0; Montoya, E.R., Terburg, D., Bos, P.A., and van Honk, J. (2012). Testosterone, cortisol, and serotonin as key regulators of social aggression: A review and theoretical perspective. *Motivation and Emotion*, *36*(1), 65–73. doi: 10.1007/s11031-011-9264-3.
19. Ross, C.P., Cyrenne, D.-L.M., and Brown, G.R. (2013). Sex differences in sensation-seeking: A meta-analysis. *Scientific Reports*, *3*, 2486. doi: 10.1038/srep02486
20. Zuckerman, M. (1985). Sensation seeking, mania, and monoamines. *Neuropsychobiology*, *13*(3), 121–128. doi: 10.1159/000118174
21. Daitzman, R.J., Zuckerman, M., Sammelwitz, P., and Ganjam, V. (2008). Sensation seeking and gonadal hormones. *Journal of Biosocial Science*, *10*(4), 401–408. doi: 10.1017/S0021932000011895
22. Carrasco, J., Sáiz-Ruiz, J., Díaz-Marsá, M., César, J., and López-Ibor, J. (1999). Low platelet monoamine oxidase activity in sensation-seeking bullfighters. *CNS Spectrums*, *4*(12), 21–24. doi: 10.1017/S1092852900006787
23. Dellu, F., Piazza, P. V, Mayo, W., Le Moal, M., and Simon, H. (1996). Novelty-seeking in rats: Biobehavioral characteristics and possible relationship with the sensation-seeking trait in man. *Neuropsychobiology*, *34*, 136–145. doi: 10.1159/000119305
24. Roberti, J.W. (2004). A review of behavioral and biological correlates of sensation seeking. *Journal of Research in Personality*, *38* (3), 256–279. doi: 10.1016/S0092-6566(03)00067-9
25. Freeman, H.D., and Beer, J.S. (2010). Frontal lobe activation mediates the relation between sensation seeking and cortisol increases. *Journal of Personality*, *78*(5), 1497–1528. doi: 10.1111/j.1467-6494.2010.00659.x

26. Lee, D., Perkins, K. A., Zimmerman, E., Robbins, G., and Kelly, T.H. (2011). Effects of 24 hours of tobacco withdrawal and subsequent tobacco smoking among low and high sensation seekers. *Nicotine & Tobacco Research: Official Journal of the Society for Research on Nicotine and Tobacco*, *13*(10), 943–54. doi: 10.1093/ntr/ntr102
27. Feij, J., and Taris, T. (2010). Beyond the genetic basis of sensation seeking: The influence of birth order, family size and parenting styles. *Romanian Journal of Applied Psychology*, *12* (2), 54–61.
28. Feij and Taris, Beyond the genetic basis of sensation seeking.
29. Feij, J.A. (1979). *Temperament: Onderzoek naar de betekenis van extraversie, emotionaliteit, impulsiviteit en spanningsbehoefte [Temper: Research on the meaning of extraversion, emotionality, impulsivity and sensation seeking]*. Lisse, The Netherlands: Swets & Zeitlinger.
30. Feij and Taris, Beyond the genetic basis of sensation seeking.
31. Boomsma, D.I., de Geus, E.J., van Baal, G.C., and Koopmans, J.R. (1999). A religious upbringing reduces the influence of genetic factors on disinhibition: Evidence for interaction between genotype and environment on personality. *Twin Research: The Official Journal of the International Society for Twin Studies*, *2*(2), 115–25. doi: 10.1375/twin.2.2.115
32. National Human Genome Research Institute (NHGRI). (2018). An Overview of the Human Genome Project, May 11. www.genome.gov /12011238/an-overview-of-the-human-genome-project/
33. Holliday, R. (2006). Epigenetics: A historical overview. *Epigenetics*, *1* (2), 76–80. doi: 10.4161/epi.1.2.2762
34. Lyko, F., Foret, S., Kucharski, R., Wolf, S., Falckenhayn, C., and Maleszka, R. (2010). The honey bee epigenomes: Differential methylation of brain DNA in queens and workers. *PLoS Biology*, *8*(11). doi: 10.1371/journal.pbio.1000506
35. Weaver, I.C., Cervoni, N., Champagne, F.A., D'Alessio, A.C., Sharma, S., Seckl, J.R., . . . and Meaney, M.J. (2004). Epigenetic programming by maternal behavior. *Nature Neuroscience*, *7*(8), 847–854. doi: 10.1038/nn1276

Chapter 3: Faster, Hotter, Louder: The Everyday Life of a High Sensation-Seeker

1. You can follow the White Rabbit's story at www .lovesharetravel.org/

2. van Brugen, I. (2018, January 13) Deal watch: Budget travel. *The Times*. www.thetimes.co.uk/edition/money/deal-watch-budget-travel-bjvh3nhgq
3. British Association for the Advancement of Science. (2011). *LaughLab: The scientific quest for the world's funniest joke*. London: Arrow Books.
4. Polimeni, J., and Reiss, J.P. (2006). The first joke: Exploring the evolutionary origins of humor. *Evolutionary Psychology*, *4*(1). doi: 10.1177/147470490600400129
5. Freud, S. (1922). *Beyond the pleasure principle*. London: International Psycho-Analytical.
6. Ruch, W. (1988). Sensation seeking and the enjoyment of structure and content of humour: Stability of findings across four samples. *Personality and Individual Differences*, *9*(5), 861–871. doi: 10.5167/uzh-77520
7. Handy, J. (1992). *Deep thoughts: Inspiration for the uninspired*. New York, NY: Berkley Books.
8. Ruch, Sensation seeking and the enjoyment of structure and content of humour.
9. Ibid.
10. Litle, P.A. (1986). *Effects of a stressful movie and music on physiological and affect arousal as a function of sensation seeking trait* (unpublished doctoral dissertation). University of Delaware, Dissertation International, 49/09B.
11. Zuckerman, M. (2006). Sensation seeking in entertainment. In J. Bryant and P. Vorderer (Eds.), *Psychology of entertainment* (pp. 367–387). Mahwah, NJ: Lawrence Erlbaum Associates.
12. Tamborini, R., Stiff, J., and Zillman, D. (1987). Preference for graphic horror featuring male versus female victimization. *Human Communication Research*, *13*(4), 529–552. doi: 10.1111/j.1468-2958.1987.tb00117.x; Slater, M.D. (2003). Alienation, aggression, and sensation seeking as predictors of adolescent use of violent film, computer, and website content. *Journal of Communication*, *53* (1), 105–121. doi: 10.1111/j.1460-2466.2003.tb03008.x; Rowland, G., Fouts, G., and Heatherton, T. (1989). Television viewing and sensation seeking: Uses, preferences and attitudes. *Personality and Individual Differences*, *10*(9), 1003–1006. doi: 10.1016/0191-8869(89)90066-4.
13. Zuckerman, M. (1979). *Sensation seeking: Beyond the optimal level of arousal*. Hillsdale, NJ: Lawrence Erlbaum Associates.

14. Brown, L.T., Ruder, V., Ruder, J., and Young, S. (1974). Stimulation seeking and the Change Seeker Index. *Journal of Consulting and Clinical Psychology*, *42*(2), 311.
15. Welsh, G.S. (1959). *Welsh Figure Preference Test*. Washington, DC: Consulting Psychologists Press.
16. Zuckerman, M., Bone, R.N., Neary, R., Mangelsdorff, D., and Brustman, B. (1972). What is the sensation seeker? Personality trait and experience correlates of the Sensation-Seeking Scales. *Journal of Consulting and Clinical Psychology*, *39*(2), 308–321. doi: 10.1037/h0033398
17. Osborne, J.W., and Farley, F.H. (1970). The relationship between aesthetic preference and visual complexity in abstract art. *Psychonomic Science*, *19*(2), 69–70.
18. Sanbonmatsu, D.M., Strayer, D.L., Medeiros-Ward, N., and Watson, J.M. (2013). Who multi-tasks and why? Multi-tasking ability, perceived multi-tasking ability, impulsivity, and sensation seeking. *PloS One*, *8*(1), e54402. http://doi.org/10.1371/journal.pone.0054402
19. Strayer, D.L., and Watson, J.M. (2012). Supertaskers and the multitasking brain. *Scientific American Mind*, *23*(1), 22–29. doi:10.1038/scientificamericanmind0312-22
20. Sanbonmatsu et al., Who multi-tasks and why?
21. Ibid.
22. Lee, J., Mehler, B., Reimer, B., and Coughlin, J.F. (2016). Sensation seeking and drivers' glance behavior while engaging in a secondary task. *Proceedings of the Human Factors and Ergonomics Society Annual Meeting*, *60* (1),1864–1868. doi: /10.1177/1541931213601425
23. Schmall, T. (2018). Vacation fantasies take up a year of our lives. *New York Post*, June 19. https://nypost.com/2018/06/19/americans-spend-nearly-a-year-of-their-lives-dreaming-about-vacation/
24. World Travel and Tourism Council (2018). *Travel & Tourism Economic Impact 2018*. www.wttc.org/-/media/files/reports/economic-impact-research/regions-2018/world2018.pdf
25. Schmall, Vacation fantasies take up a year of our lives.
26. Gilchrist, H., Povey, R., Dickinson, A., and Povey, R. (1995). The sensation seeking scale: Its use in a study of the characteristics of people choosing "adventure holidays." *Personality and Individual Differences*, *19*(4), 513–516. doi: 10.1016/0191-8869(95)00095-N

27. Franken, R.E., Gibson, K.J., and Rowland, G.L. (1992). Sensation seeking and the tendency to view the world as threatening. *Personality and Individual Differences*, *13*(1), 31–38. doi: 10.1016/0191-8869(92)90214-A
28. Pizam, A., Jeong, G., Reichel, A., van Boemmel, H., Lusson, J.M., Steynberg, L., . . . Montmany, N. (2004). The relationship between risk-taking, sensation-seeking, and the tourist behavior of young adults: A cross-cultural study. *Journal of Travel Research*, *42*(3), 251–260. doi: 10.1177/0047287503258837
29. Lepp, A., and Gibson, H. (2008). Sensation seeking and tourism: Tourist role, perception of risk and destination choice. *Tourism Management*, *29*, 740–750. doi: 10.1016/j.tourman.2007.08.002
30. Bourdain, A. (2013). *Kitchen confidential.* London: A&C Black; Bourdain, A. (2001). *Typhoid Mary: An urban historical.* London: Bloomsbury; Bourdain, A. (2010). *A cook's tour: In search of the perfect meal.* London: Bloomsbury;Bourdain, A. (2007). *No reservations: Around the world on an empty stomach.* London: Bloomsbury.
31. Caseras, F.X., Fullana, M.A., Riba, J., Barbanoj, M.J., Aluja, A., and Torrubia, R. (2006). Influence of individual differences in the Behavioral Inhibition System and stimulus content (fear versus blood-disgust) on affective startle reflex modulation. *Biological Psychology*, *72*(3), 251–256. doi: 10.1016/j.biopsycho.2005.10.009
32. Dvorak, R.D., Simons, J.S., and Wray, T.B. (2011). Poor control strengthens the association between sensation seeking and disgust reactions. *Journal of Individual Differences*, *32*(4), 219–224. doi: 10.1027/1614-0001/a000054
33. Byrnes, N.K., and Hayes, J.E. (2016). Behavioral measures of risk tasking, sensation seeking and sensitivity to reward may reflect different motivations for spicy food liking and consumption. *Appetite*, *103*, 411–422. doi:10.1016/J.APPET.2016.04.037
34. Zajonc, R.B., and Markus, H. (1982). Affective and cognitive factors in preferences. *Journal of Consumer Research*, *9*(2), 123–131.
35. Take a look at their blog for more information, www.crafthotsauce.com/blog/2018-hot-sauce-and-chili-festivals
36. Byrnes, N.K., and Hayes, J.E. (2016). Behavioral measures of risk tasking, sensation seeking and sensitivity to reward may reflect

different motivations for spicy food liking and consumption. *Appetite*, *103*, 411–422. http://doi.org/10.1016/J.APPET.2016.04.037

37. Guinness World Records. (2017). Hottest chilli pepper. August 11. www.guinnessworldrecords.com/world-records/hottest-chili/

Chapter 4: Lights, Camera, Action: Sports and Adventure in High Sensation-Seeking

1. Davis, M. (2014). *Down and dirty: The essential training guide for obstacle races and mud runs*. Beverly, MA: Fair Winds Press.
2. Ibid.
3. Perez, A. (2015). Obstacle races going mainstream, more popular than marathons. *USA Today*, November 2. www.usatoday.com/story/sports/2015/11/02/obstacle-races-going-mainstream-more-popular-than-marathons/73743474/
4. Zuckerman, M. (1996). Sensation seeking and the taste for vicarious horror. In J.B. Weaver and R. Tamborini (Eds.), *Horror films: Current research on audience preferences and reactions* (pp. 147–160). Hoboken, NJ: Taylor & Francis.
5. Greenberg, M.R., Kim, P.H., Duprey, R.T., Jayant, D.A., Steinweg, B. H., Preiss, B.R., and Barr Jr., G.C. (2014). Unique obstacle race injuries at an extreme sports event: A case series. *Annals of Emergency Medicine*, *63*(3), 361–366. doi: 10.1016/j.annemergmed.2013.10.008
6. Cronin, C. (1991). Note and shorter communication: Sensation seeking among mountain climbers. *Personality and Individual Differences*, *12*(6), 653–654.
7. Hymbaugh, K., and Garrett, J. (1974). Sensation seeking among skydivers. *Perceptual and Motor Skills*, *38*(1), 118.
8. Straub, W.F. (1982). Sensation seeking among high and low-risk male athletes. *Journal of Sport Psychology*, 4, 246–253.
9. Cronin, Note and shorter communication.
10. Jack, S.J., and Ronan, K.R. (1998). Sensation seeking among high- and low-risk sports participants. *Personality and Individual differences*, *25*(6), 1063–1083.
11. Breivik, G., Sand, T.S., and Sookermany, A.M.D. (2017). Sensation seeking and risk-taking in the Norwegian population. *Personality and Individual Differences*, *119*, 266–272. doi: 10.1016/j.paid.2017.07.039

12. Smith, R.E., Ptacek, J.T., and Smoll, F.L. (1992). Sensation seeking, stress, and adolescent injuries: A test of stress buffering, risk-taking, and coping skills hypotheses. *Journal of Personality and Social Psychology*, *62*(6), 1016–1024.
13. Ditunno, P.L., McCauley, C., and Marquette, C. (1985). Sensation-seeking behavior and the incidence of spinal cord injury. *Archives of Physical Medicine and Rehabilitation*, *66*(3), 152–155.
14. Zuckerman, M. (1994). *Behavioral expressions and biosocial bases of sensation seeking*. Cambridge: Cambridge University Press.
15. Ibid.
16. Gamble, T., and Walker, I. (2016). Wearing a bicycle helmet can increase risk taking and sensation seeking in adults. *Psychological Science*, *27*(2), 289–294. doi: 10.1177/0956797615620784
17. Adams, J., and Hillman, M. (2001). The risk compensation theory and bicycle helmets. *Injury Prevention*, *7*(2), 89–91. doi: 10.1136/ip.7.2.89
18. Sagberg, F., Fosser, S., and Saetermo, I.A. (1997). An investigation of behavioural adaptation to airbags and antilock brakes among taxi drivers. *Accident; Analysis and Prevention*, *29*(3), 293–302.
19. Morrongiello, B.A., Lasenby, J., and Walpole, B., (2007). Risk compensation in children: Why do children show it in reaction to wearing safety gear? *Journal of Applied Developmental Psychology*, *28*(1), 56–63. doi: 10.1016/j.appdev.2006.10.005
20. Phillips, R.O., Fyhri, A., and Sagberg, F. (2011). Risk compensation and bicycle helmets. *Risk Analysis: An International Journal*, *31*(8), 1187–1195.
21. Liptak, A. (2018). British Tesla driver banned after caught in the passenger seat while Autopilot was engaged. www.theverge.com/2018/4/29/17298750/tesla-autopilot-british-driver-charged-driving-sleeping
22. Williams, F. (2014). The science of conquering your greatest fears. *Outside Magazine*. www.outsideonline.com/1926246/science-conquering-your-greatest-fears
23. Zuckerman, M. (1979). *Sensation seeking: Beyond the optimal level of arousal*. Hillsdale, NJ: Lawrence Erlbaum Associates.
24. Science North (2018). About Us. *Sciencenorth.ca*. www.sciencenorth.ca/about/

25. Thank you to Amy Wilson, Will Gadd, Katherine Beatlie, Jeb Corliss, and the staff of Science North for providing me with access to the interviews.

26. Strauss, G. (2011). Will Gadd is fearless for Discovery's newest 'Planet'. *USA Today*, March 23. https://usatoday30.usatoday.com/printedition/life/20071109/wk_fearless09.art.htm
27. City of Niagara Falls. (2018). Niagara Falls facts. *Nigara Falls Canada*, September 17. https://niagarafalls.ca/living/about-niagara-falls/facts.aspx
28. Gadd, W. (2018). *Will Gadd the athlete*, September 17. http://willgadd.com/will-gadd-the-athlete/
29. Immonen, T., Brymer, E., Davids, K., Liukkonen, J., and Jaakkola, T. (2018). An ecological conceptualization of extreme sports. *Frontiers in Psychology*, *9*. doi: 10.3389/fpsyg.2018.01274
30. Duckworth, A.L., Peterson, C., Matthews, M.D., and Kelly, D.R. (2007). Grit: Perseverance and passion for long-term goals. *Journal of Personality and Social Psychology*, *92*(6), 1087–1101. doi: 10.1037/0022-3514.92.6.1087
31. Duckworth, A. (2018). Grit: The power of passion and perseverance [Video file]. www.ted.com/talks/angela_lee_duckworth_grit_the_power_of_passion_and_perseverance
32. Duckworth, A. (2018). Angela Duckworth. http://angeladuckworth.com/qa/
33. Rimfeld, K., Kovas, Y., Dale, P.S., and Plomin, R. (2016). True grit and genetics: Predicting academic achievement from personality. *Journal of Personality and Social Psychology*, *111*(5), 780–789. doi: 10.1037/pspp0000089
34. Credé, M., Tynan, M.C., and Harms, P.D. (2017). Much ado about grit: A meta-analytic synthesis of the grit literature. *Journal of Personality and Social Psychology*, *113*(3), 492.
35. Pew Research Center (2018). Social Media Use 2018. *Pew Research Center: Internet, Science & Tech*. www.pewinternet.org/2018/03/01/social-media-use-in-2018/
36. Soat, M. (2015). Social media triggers a dopamine high. *Marketing News*, *49*(11), 20–21.
37. @jebcorliss Instagram photos and videos. (2018). www.instagram.com/jebcorliss/?hl=en
38. Barlow, M., Woodman, T., and Hardy, L. (2013). Great expectations: Different high-risk activities satisfy different motives. *Journal of Personality and Social Psychology*, *105*(3), 458–75. doi: 10.1037/a0033542
39. Ospina, M.B., Bond, K., Karkhaneh, M., Tjosvold, L., Vandermeer, B., Liang, Y., … Klassen, T.P. (2007). Meditation

practices for health: State of the research. *Evidence report/ technology assessment*, *155*, 1–263.

40. Barlow et al., Great expectations.

Chapter 5: What About Your Friends? The Relationships of High Sensation-Seekers.

1. Zuckerman, M. (1979). *Sensation seeking: Beyond the optimal level of arousal.* Hillsdale, NJ: Lawrence Erlbaum Associates.
2. Franken, R.E., Gibson, K.J., and Mohan, P. (1990). Sensation seeking and disclosure to close and casual friends. *Personality and Individual Differences*, *11*(8), 829–832. doi: 10.1016/0191-8869(90)90192-T
3. Zuckerman, Sensation seeking.
4. Franken et al., Sensation seeking and disclosure to close and casual friends.
5. McKay, S., Skues, J.L., and Williams, B.J. (2018). With risk may come reward: Sensation seeking supports resilience through effective coping. *Personality and Individual Differences*, *121*, 100–105. doi:10.1016/j.paid.2017.09.030
6. Fiske, S.T. (2003). Five core social motives, plus or minus five. *Motivated Social Perception*, *9*, 233–246.
7. Thornton, B., Ryckman, R., and Gold, J. (1981). Sensation seeking as a determinant of interpersonal attraction toward similar and dissimilar others. *Journal of Mind and Behavior*, *2*(1), 85–91.
8. Weisskirch, R.S., and Murphy, L.C. (2004). Friends, porn, and punk: Sensation seeking in personal relationships, internet activities, and music preference among college students. *Adolescence*, *39*(154), 189–201.
9. Perlman, D., and Fehr, B. (1986). Theories of friendship: The analysis of interpersonal attraction. In S. Duck and D. Perlman (Eds.), *Friendship and social interaction* (pp. 9–40). New York, NY: Springer.
10. Thornton et al., Sensation seeking as a determinant of interpersonal attraction.
11. Swissotel Hotels & Resorts. (2018). The Ultimate Guide to Worldwide Etiquette. www.swissotel.com/promo/etiquette-map/
12. Thornton et al., Sensation seeking as a determinant of interpersonal attraction.

13. Arasaratnam, L.A., and Banerjee, S.C. (2011). Sensation seeking and intercultural communication competence: A model test. *International Journal of Intercultural Relations, 35*(2), 226–233. doi:10.1016/j.ijintrel.2010.07.003
14. Gareis, E. (1995). *Intercultural friendship: A qualitative study.* Lanham, MD: University Press of America; Stanley Budner, N.Y. (1962). Intolerance of ambiguity as a personality variable 1. *Journal of Personality, 30*(1), 29–50.
15. Morgan, S., and Arasaratnam, L.A. (2003). Intercultural friendships as social excitation: Sensation seeking as a predictor of intercultural friendship seeking behavior. *Journal of Intercultural Communication Research, 32*, 175–186.
16. Spencer-Rodgers, J., and McGovern, T. (2002). Attitudes toward the culturally different: The role of intercultural communication barriers, affective responses, consensual stereotypes, and perceived threat. *International Journal of Intercultural Relations, 26*(6), 609–631. doi: /10.1016/S0147-1767(02)00038-X
17. Lee, J. (1973). *Colours of love: An exploration of the ways of loving.* New York, NY: New Press.
18. Woll, S.B. (1989). Personality and relationship correlates of loving styles. *Journal of Research in Personality, 23*(4), 480–505.
19. Franken, R.E. (1993). Sensation seeking and keeping your options open. *Personality and Individual Differences, 14*(1), 247–249. doi:10.1016/0191-8869(93)90196-A
20. Hoyle, R.H., Fejfar, M.C., and Miller, J.D. (2000). Personality and sexual risk taking: A quantitative review. *Journal of Personality, 68*(6), 1203–1231.
21. Curry, I., Luk, J.W., Trim, R.S., Hopfer, C.J., Hewitt, J.K., Stallings, M.C., ... Wall, T.L. (2018). Impulsivity dimensions and risky sex behaviors in an at-risk young adult sample. *Archives of Sexual Behavior, 47*(2), 529–536. http://doi.org/10.1007/s10508-017-1054-x
22. Hoyle et al. (2000). Personality and sexual risk taking.
23. Salovey, P., and Mayer, J.D. (1990). Emotional intelligence. *Imagination, Cognition and Personality, 9*(3), 185–211.
24. Goleman, D. (2006). *Emotional intelligence.* New York, NY: Bantam Books.
25. Bacon, A., Burak, H., and Rann, J. (2014). Sex differences in the relationship between sensation seeking, trait emotional intelligence and delinquent behaviour. *The Journal of Forensic*

Psychiatry & Psychology, *25*(6), 673–683. doi: 10.1080/14789949.2014.943796

26. Roberti, J.W. (2004). A review of behavioral and biological correlates of sensation seeking. *Journal of Research in Personality*, *38*(3), 256–279. http://doi.org/10.1016/S0092-6566(03)00067-9

Chapter 6: All in a Day's Work

1. Musolino, R.F., and Hershenson, D.B. (1977). Avocational sensation seeking in high and low risk-taking occupations. *Journal of Vocational Behavior*, *10*(3), 358–365. doi: 10.1016/0001-8791(77)90069-0
2. Waters, C.W., Ambler, R., and Waters, L.K. (1976). Novelty and sensation seeking in two academic training settings. *Educational and Psychological Measurement*, *36*(2), 453–457. doi: 10.1177/001316447603600227
3. Zaleski, Z. (1984). Sensation-seeking and risk-taking behaviour. *Personality and Individual Differences*, *5*(5), 607–608. doi: 10.1016/0191-8869(84)90039-4
4. Levin, B.H., and Brown, W.E. (1975). Susceptibility to boredom of jailers and law enforcement officers. *Psychological Reports*, *36*(1), 190. doi: 10.2466/pr0.1975.36.1.190
5. Golding, J.F., and Cornish, A.M. (1987). Personality and life-style in medical students: Psychopharmacological aspects. *Psychology and Health*, *1*, 287–301.
6. Mustika, M.D., and Jackson, C.J. (2016). How nurses who are sensation seekers justify their unsafe behaviors. *Personality and Individual Differences*, *100*, 79–84. doi: 10.1016/j.paid.2016.02.008
7. Ibid.
8. Jones, T. (2016). Ask the Astronaut: Is it quiet onboard the space station? *Air & Space Magazine*, April 27. www.airspacemag.com/ask-astronaut/ask-astronaut-it-quiet-onboard-space-station-180958932/#2sBkQOw0XWJUIIvR.99
9. Lazzaro, S. (2017). Astronaut Scott Kelly reveals the ISS smells "like JAIL." *Daily Mail*, September 6. www.dailymail.co.uk/sciencetech/article-4859226/Astronaut-Scott-Kelly-reveals-ISS-smells-like-JAIL.html
10. Migneault, J. (2016). Is this Science North scientist Canada's next astronaut? *Sudbury.com*, December 19. www.sudbury.com/local-

news/is-this-science-north-scientist-canadas-next-astronaut-492993

11. CBC. (2016). Meet the Sudbury scientist who's shooting for the stars. December 15. www.cbc.ca/news/canada/sudbury/sudbury-scientist-next-astronaut-1.3898357
12. Sunder, J., Sunder, S.V., and Zhang, J. (2017). Pilot CEOs and corporate innovation. *Journal of Financial Economics*, *123*(1), 209–224. doi: https://doi.org/10.1016/j.jfineco.2016.11.002
13. Nair, P. (2017). RESEARCH: "Sensation seekers" make better leaders. *Growth Business*, February 22. www.growthbusiness.co.uk/research-sensation-seekers-make-better-leaders-2549779/

Chapter 7: The Dark Side of High Sensation-Seeking

1. Zuckerman, M. (1994). *Behavioral expressions and biosocial bases of sensation seeking*. Cambridge: Cambridge University Press, 27.
2. Zuckerman, M., and Neeb, M. (1979). Sensation seeking and psychopathology. *Psychiatry Research*, *1*(3), 255–264.
3. Robertson, L. (2018). Road death trend in the United States: Implied effects of prevention. *Journal of Public Health Policy*, *39*(2), 193–202. doi: 10.1057/s41271-018-0123-2
4. Wilde, G. (1982). The theory of risk homeostasis: Implications for safety and health. *Risk Analysis*, *2*, 209–225.
5. Naatanen, R., and Summala, H. (1974). A model for the role of motivational factors in drivers' decision making. *Accident Analysis and Prevention*, *6*, 243–261.
6. Thiffault, P., and Bergeron, J. (2003). Fatigue and individual differences in monotonous simulated driving. *Personality and Individual Differences*, *34*(1), 159–176. doi: 0.1016/S0191-8869(02)00119-8
7. Campbell, M., and Stradling, S. (2003). Factors influencing driver speed choices. In *Behavioural Research in Road Safety XII*. London: Department for Transport.
8. Peer, E., and Rosenbloom, T. (2013). When two motivations race: The effects of time-saving bias and sensation-seeking on driving speed choices. *Accident Analysis and Prevention*, *50*, 1135–1139. doi: 10.1016/j.aap.2012.09.002
9. Ibid.

10. Heino, A., van den Molen, H.H., and Wilde, G.J.S. (1992). Risk homeostasis process in car following and perceived risk. Report VK 92-02 Rijksuniversiteit Groningen Haven, The Netherlands.
11. Furnham, A., and Saipe, J. (1993). A conceptualization of driving behavior as threat avoidance. *Ergonomics*, *27*, 1139–1155.
12. Sanbonmatsu, D.M., Strayer, D.L., Medeiros-Ward, N., and Watson, J.M. (2013). Who multi-tasks and why? Multi-tasking ability, perceived multi-tasking ability, impulsivity, and sensation seeking. *PloS One*, *8*(1), e54402. doi: 10.1371/journal.pone.0054402
13. Schwebel, D.C., Severson, J., Ball, K.K., and Rizzo, M. (2006). Individual difference factors in risky driving: The roles of anger/hostility, conscientiousness, and sensation-seeking. *Accident Analysis & Prevention*, *38*(4), 801–810. doi: 10.1016/j.aap.2006.02.004
14. Wilson, R.J. (1990). The relationship of seat belt non-use to personality, lifestyle and driving record. *Health Education Research*, *5* (2), 175–185. doi: 10.1093/her/5.2.175
15. Witte, K., and Donohue, W.A. (2000). Preventing vehicle crashes with trains at grade crossings: The risk seeker challenge. *Accident Analysis & Prevention*, *32*(1), 127–139.
16. Jonah, B.A. (1997). Sensation seeking and risky driving: A review and synthesis of the literature. *Accident Analysis & Prevention*, *29*(5), 651–665. doi: 10.1016/s0001-4575(97)00017-1
17. Lang, A.R. (1983). Addictive personality: A viable construct? In P.K. Levison, D.R. Gerstein, and D.R. Maloff (Eds.), *Commonalities in substance abuse and habitual behavior* (pp. 157–236). Lanham, MD: Lexington Books; Walters, S.T., and Rotgers, F. (Eds.). (2011). *Treating substance abuse: Theory and technique*. New York: Guilford Press.
18. Segal, B. (1975). Personality factors related to drug and alcohol use. In D. Lettieri (Ed.), *Predicting adolescent drug abuse: A review of issues, methods, and correlates* (pp. 165–191). Rockville, MD: National Institute on Drug Abuse.
19. Dubey, C., and Arora, M. (2008). Sensation seeking level and drug of choice. *Journal of the Indian Academy of Applied Psychology*, *34*(1), 73–82.
20. Kopstein, A., Crum, R., and Celentano, D. (2001). Sensation seeking needs among 8th and 11th graders: Characteristics associated with cigarette and marijuana use. *Drug and Alcohol Dependence*, *62*(3), 195–203.

21. Zuckerman, M. (1979). *Sensation seeking: Beyond the optimal level of arousal*. Hillsdale, NJ: Lawrence Erlbaum Associates.
22. Gorsuch, R.L., and Butler, M.C. (1976). Initial drug abuse: A review of predisposing social psychological factors. *Psychological Bulletin*, *83*(1), 120.
23. Cunningham-Williams, R.M., and Cottler, L.B. (2001). The epidemiology of pathological gambling. *Seminars in Clinical Neuropsychiatry*, *6*(3), 155–166.
24. Harlow, W., and Brown, K. (1990). *The role of risk tolerance in the asset allocation process: A new perspective*. Charlottesville, VA: Research Foundation of the Institute of Chartered Financial Analysts.
25. Kuley, N.B., and Jacobs, D.F. (1988). The relationship between dissociative-like experiences and sensation seeking among social and problem gamblers. *Journal of Gambling Behavior*, *4*(3), 197–207. doi:10.1007/BF01018332
26. American Psychiatric Association. (2013). *Diagnostic and statistical manual of mental disorders: DSM-5* (5th edn). Washington, DC: American Psychiatric Association.
27. McDaniel, S.R., and Zuckerman, M. (2003). The relationship of impulsive sensation seeking and gender to interest and participation in gambling activities. *Personality and Individual Differences*, *35*(6), 1385–1400. doi:10.1016/S0191-8869(02)00357-4
28. Alessi, S.M., and Petry, N.M. (2003). Pathological gambling severity is associated with impulsivity in a delay discounting procedure. *Behavioural Processes*, *64*(3), 345–354.
29. American Psychiatric Association, Diagnostic and statistical manual of mental disorders.
30. Amini, T. (2015). Peeing in a coffee can, and other gaming marathon stories. *Kotaku.com*, June 12. https://kotaku.com/peeing-in-a-coffee-can-and-other-gaming-marathon-stori-1710453756
31. Kemp. J. (2013). Oklahoma parents so engulfed in Second Life they allegedly starved their real 3-year-old daughter. *New York Daily News*, October 12. www.nydailynews.com/news/national/oklahoma-parents-engulfed-online-fantasy-world-allegedly-starved-real-3-year-old-daughter-cops-article-1.1483479
32. Wang, J., Chen, J., Yang, L., and Gao, S. (2013). Meta-analysis of the relationship between sensation seeking and internet addiction. *Advances in Psychological Science*, *21*(10), 1720–1730. doi:10.3724/SP.J.1042.2013.01720

33. Gilbert, M.A., and Bushman, B.J. (2017). Frustration-aggression hypothesis. In *Encyclopedia of Personality and Individual Differences* (pp. 1–3). Cham: Springer. /doi.org/10.1007/978-3-319-28099-8_816-1
34. Gottfredson, M.R., and Hirschi, T. (1990). *A general theory of crime.* Palo Alto, CA: Stanford University Press.
35. Joireman, J., Anderson, J., and Strathman, A. (2003). The aggression paradox: Understanding links among aggression, sensation seeking, and the consideration of future consequences. *Journal of Personality and Social Psychology, 84*(6), 1287–1302. doi:10.1037/0022-3514.84.6.1287
36. Ibid.
37. Allen, J.J., Anderson, C.A., and Bushman, B.J. (2018). The General Aggression Model. *Current Opinion in Psychology, 19,* 75–80. doi: 10.1016/J.COPSYC.2017.03.034
38. Joireman et al., The aggression paradox.
39. Simó, S., and Pérez, J. (1991). Sensation seeking and antisocial behaviour in a junior student sample. *Personality and Individual Differences, 12*(9), 965–966. doi:10.1016/0191-8869(91)90186-F
40. Pfefferbaum, B., and Wood, P.B. (1994). Self-report study of impulsive and delinquent behavior in college students. *The Journal of Adolescent Health: Official Publication of the Society for Adolescent Medicine, 15*(4), 295–302.
41. Mischel, W. (2014). *The marshmallow test: Mastering self-control.* New York: Little, Brown and Company.

Chapter 8: Super Power or Super Problem

1. Thoreau, H.D. (1908). *Walden, or, life in the woods.* London: J.M. Dent.
2. Miller, L., Spiegel, A., and Rosin, H. (2015). *Fearless* [Audio podcast], January 16. www.npr.org/programs/invisibilia/377515477/fearless
3. DiGiandomenico, S., Masi, R., Cassandrini, D., El-Hachem, M., DeVito, R., Bruno, C., and Santorelli, F.M. (2006). Lipoid proteinosis: Case report and review of the literature. *Acta Otorhinolaryngol Ital, 26,* 162–167.
4. Norbury, A., and Husain, M. (2015). Sensation-seeking: Dopaminergic modulation and risk for psychopathology. *Behavioural Brain Research, 288,* 79–93. doi: 10.1016/j.bbr.2015.04.015

5. Thornton, N. (2012). MV *Herald of Free Enterprise* - past and present, May 22. www.doverferryphotosforums.co.uk/mv-herald-of-free-enterprise-past-and-present/
6. Joseph, S.A., Brewin, C.R., Yule, W., and Williams, R. (1991). Causal attributions and psychiatric symptoms in survivors of the *Herald of Free Enterprise* disaster. *The British Journal of Psychiatry*, *159*(4), 142–146. doi:10.1192/bjp.159.4.542
7. American Psychiatric Association. (2013). *Diagnostic and statistical manual of mental disorders* (5th edn). Arlington, VA: American Psychiatric Publishing.
8. Lusk, J.D., Sadeh, N., Wolf, E.J., and Miller, M.W. (2017). Reckless self-destructive behavior and PTSD in veterans: The mediating role of new adverse events. *Journal of Traumatic Stress*, *30*(3), 270–278. doi: 10.1002/jts.22182
9. Horrom, T. (2018). Reckless behavior fuels ongoing stress for some with PTSD. *Research.va.gov*, June 14. www.research.va.gov/currents/0617-Reckless_behavior_fuels_ongoing_stress_PTSD.cfm
10. Orr, S.P., Claiborn, J.M., Altman, B., Forgue, D.F., de Jong, J.B., Pitman, R.K., and Herz, L.R. (1990). Psychometric profile of posttraumatic stress disorder, anxious, and healthy Vietnam veterans: Correlations with psychophysiologic responses. *Journal of Consulting and Clinical Psychology*, *58*(3), 329–335.
11. Matthews, G., and Wells, A. (2000). Attention, automaticity, and affective disorder. *Behavior Modification*, *24*(1), 69–93. doi:10.1177/0145445500241004
12. Grossman, D. (2007). *On combat* (3rd edn). Millstadt, IL: Warrior Science Publications.
13. Wymer, W., Self, D., and Findley, C.S. (2008). Sensation seekers and civic participation: Exploring the influence of sensation seeking and gender on intention to lead and volunteer. *International Journal of Nonprofit and Voluntary Sector Marketing*, 13(4), 287–300. doi: 10.1002/nvsm.330
14. Smith, S.F., Lilienfeld, S.O., Coffey, K., and Dabbs, J.M. (2013). Are psychopaths and heroes twigs off the same branch? Evidence from college, community, and presidential samples. *Journal of Research in Personality*, 47(5), 634–646. doi: 10.1016/j.jrp.2013.05.006
15. Patton, C.L., Smith, S.F., and Lilienfeld, S.O. (2018). Psychopathy and heroism in first responders: Traits cut from the same cloth? *Personality Disorders: Theory, Research, and Treatment*, 9(4), 354–368. http://dx.doi.org/10.1037/per0000261

16. Farley, F. (2012). The real heroes of "The Dark Knight" Big H heroism. *Psychology Today*. www.psychologytoday.com/us/blog/the-peoples-professor/201207/the-real-heroes-the-dark-knight
17. Ibid.
18. For more information about Paradox Sports, see their website www.paradoxsports.org/
19. Eschleman, K.J., Bowling, N.A., and Alarcon, G.M. (2010). A meta-analytic examination of hardiness. *International Journal of Stress Management*, *17*(4), 277. doi: 10.1037/a0020476
20. Maddi, S.R. (2006). Hardiness: The courage to grow from stresses. *The Journal of Positive Psychology*, *1*(3), 160–168. doi: 10.1080/17439760600619609
21. McKay, S., Skues, J.L., and Williams, B.J. (2018). With risk may come reward: Sensation seeking supports resilience through effective coping. *Personality and Individual Differences*, *121*, 100–105. doi:10.1016/j.paid.2017.09.030
22. Krznaric, R. (2012). Six habits of highly empathic people. *Greater Good Magazine*, November 27. https://greatergood.berkeley.edu/article/item/six_habits_of_highly_empathic_people1

Conclusion

1. Stellar, J.E., John-Henderson, N., Anderson, C.L., Gordon, A.M., McNeil, G.D., and Keltner, D. (2015). Positive affect and markers of inflammation: Discrete positive emotions predict lower levels of inflammatory cytokines. *Emotion*, *15*(2), 129–133. doi: 10.1037/emo0000033
2. Csikszentmihalyi, M. (1990). *Flow: The psychology of optimal experience*. New York, NY: Harper and Row.

INDEX